Design for Additive Manufacturing

Fundamental Principles and Advanced Strategies

Yaoyao Fiona Zhao
McGill University

Yunlong Tang
Monash University

Sheng Yang
University of Guelph

Guoying Dong
University of Colorado Denver

WILEY

Published by John Wiley & Sons, Inc., Hoboken, New Jersey.

For general information on our other products and services or for technical support, please contact our Customer Care Department within the United States at (800) 762-2974, outside the United States at (317) 572-3993 or fax (317) 572-4002.

Wiley also publishes its books in a variety of electronic formats. Some content that appears in print may not be available in electronic formats. For more information about Wiley products, visit our web site at www.wiley.com.

Library of Congress Cataloging-in-Publication Data has been applied for:

Hardback ISBN: 9781394168408
ePDF ISBN: 9781394168422
ePub ISBN: 9781394168415
oBook ISBN: 9781394168439

Cover Design: Wiley
Cover Image: © Pixel B/Shutterstock

Dedicated to the students, researchers, and practitioners advancing the field of design for additive manufacturing. This book is but one step forward, built upon the foundations laid by the many pioneers and scholars whose contributions continue to guide and inspire us. We hope it will, in turn, inspire future engineers and researchers to push the boundaries of design and manufacturing even further.

Table of Contents

guide early-stage decision-making. Chapter 4 addresses part consolidation and assembly design, providing systematic rules and functional-entity–based methods, supported by case studies.

Chapter 5 focuses on the restrictive dimensions of DfAM. It introduces general manufacturing constraints common across AM processes, followed by process-specific guidelines for polymers and metals, and concludes with an integrated presentation of design rules across AM technologies. Chapter 6 explores the opportunistic approaches of DfAM, focusing on optimization strategies such as topology optimization and multi-functional design, illustrated through selected case studies. Chapter 7 builds on this by examining the design and optimization of lattice structures fabricated via AM. Chapter 8 then transitions to advanced topics in DfAM, including digital materials, voxel-based printing, and the design of continuous fiber-reinforced composites. These emerging paradigms emphasize the integration of computational and machine learning methods to achieve unprecedented levels of product performance and efficiency. Finally, Chapter 9 reviews the latest developments in data-driven and machine learning-enabled DfAM, concluding with current limitations and outlining future research opportunities in the field.

This book is intended for researchers, graduate students, and advanced practitioners in design and manufacturing who seek both foundational knowledge and advanced methods in DfAM. It is equally relevant to industry professionals seeking structured guidance on incorporating AM into their design practices. While each chapter can be read independently, the book as a whole provides a progressive journey from established design principles to advanced multifunctional opportunities, reflecting the evolving landscape of additive manufacturing design.

It is our hope that this book will serve as both a reference and a source of inspiration, supporting the education of future engineers and the advancement of design practice in industry and academia. As AM continues to develop, so will the opportunities for design innovation. The methods and examples presented here are intended to provide a rigorous yet forward-looking foundation for that ongoing evolution.

Acknowledgments

The development of this book has been made possible through the contributions and support of many individuals and institutions.

We would first like to acknowledge the financial support that enabled much of the research presented in this book. The Natural Sciences and Engineering Research Council of Canada (NSERC), including the Discovery Grant and Discovery Accelerator Supplement, provided essential funding that allowed us to explore fundamental questions in design for additive manufacturing. Additional support from McGill University, including the William Dawson Fellowship and the McGill Engineering Doctoral Scholarship (MEDA), further strengthened this research program.

This book draws heavily from the doctoral research of the co-authors, which pioneered methods in different aspects of DfAM. We extend our gratitude to the many undergraduate students, graduate students, postdoctoral researchers, and research staff whose dedicated work contributed to the case studies, methods, and examples featured here. Their creativity, persistence, and rigor have been indispensable.

We would also like to thank our colleagues in the design and manufacturing research community. Numerous discussions at conferences, workshops, and seminars have shaped our perspectives and pushed us to refine our approaches. The opportunities to serve within the ASME Design Automation community, Design for Manufacturing and the Life Cycle community, the Design Society, and related professional organizations have been invaluable in broadening the impact of this work.

Finally, each of the authors wishes to acknowledge the encouragement and patience of their families. Their understanding and unwavering support have sustained us throughout the many years of research and writing that culminated in this book.

Yaoyao Fiona Zhao

The co-authors of this book were the first three PhD students of my academic career. Having such talented, driven, and hardworking students was the greatest gift a junior faculty member could hope for, and I consider myself fortunate to have had the opportunity to work with them. Their doctoral research not only contributed to this book but also shaped my growth as a researcher and mentor. During the years in which this work was developed, my two sons were born, and I transitioned from a junior faculty member to a tenured professor. These research endeavors are therefore deeply meaningful, marking both professional milestones and personal journeys. I am profoundly grateful to my children, my husband, and my parents for their unwavering love and support. To my sons, Ernest and Leon, you are my constant source of strength and inspiration.

Yunlong Tang

The co-author of this book, Professor Yaoyao Fiona Zhao, was also my PhD supervisor. Her guidance and mentorship have been instrumental in shaping my research career and leading me onto the path of Design for Additive Manufacturing. Under her supervision, I not only gained the foundations of rigorous research but also grew into an independent researcher capable of pursuing my own ideas and directions. Working with her has been an immense privilege, and I remain deeply grateful for the opportunities and insights she has given me.

I am equally indebted to my family for their unwavering love and support throughout this journey. To my daughter, Tina, and my son, Eric, you are my constant source of strength and inspiration. To my wife, Qianqian, you have always been my strongest support, and to my parents, Jiancheng and Jinghua, your encouragement and belief in me have been invaluable. This book is as much a reflection of your support as it is of my work.

Sheng Yang

I would like to express my deepest gratitude to my supervisor, Professor Yaoyao Fiona Zhao, whose guidance and support throughout my PhD and postdoctoral studies have been invaluable. Her mentorship has shaped me into an independent researcher, supervisor, and educator, and her encouragement has left a lasting impact on my academic journey. I am also deeply appreciative of the opportunities and resources provided by McGill University, which created an exceptional environment for my growth as a scholar.

My sincere thanks extend to my collaborators and colleagues at ADML, whose expertise, generosity, and shared commitment have enriched my work and broadened my horizons. I also treasure the memorable experiences and friendships that have made this journey both meaningful and unforgettable. Finally, I owe my deepest gratitude to my family—my wife—Bo, my daughters—Gracelyn and Katelyn, and my parents—Meihua and Shengfa—whose love, patience, and unwavering support have been the foundation of all my endeavors. This book is dedicated to them.

Guoying Dong

I am deeply grateful to my Ph.D. advisor and co-author, Professor Yaoyao Fiona Zhao, for her invaluable guidance, encouragement, and mentorship throughout my doctoral studies and beyond. Her guidance not only shaped the direction of my research but also influenced the way I think as a scholar. I am also grateful to my colleagues at ADML, especially my co-authors, Yunlong and Sheng, whose generosity, encouragement, and collaboration have been invaluable throughout this journey. As senior Ph.D. students when I began my own studies, they provided not only academic guidance but also personal support, which greatly helped me navigate my research path.

I am profoundly thankful to my family—my wife, my son, and my parents—for their love, patience, and support throughout this journey. My wife, Lei, has been my constant companion, offering support and inspiration every step of the way. During the preparation of this book, my son, Lucas, came into the world, filling my life with joy, motivation, and a renewed sense of purpose. My parents, Fengxia and Jindou, have devoted all their love to me, and their belief in me has been a constant source of strength and motivation. Without their sacrifices, encouragement, and enduring faith, this book would not have been possible.

List of Abbreviations

3MF	3D Manufacturing Format
ABS	Acrylonitrile Butadiene Styrene
ACDAM	Automated Candidacy Detection for Additive Manufacturing
AD	Axiomatic Design
AFP	Automated Fiber Placement
AI	Artificial Intelligence
AM	Additive Manufacturing
AMF	Additive Manufacturing File
AMK	Additive Manufacturing Knowledge
API	Application Programming Interface
ASME	American Society of Mechanical Engineers
ASTM	American Society for Testing and Materials
BDTR	Boosted Decision Tree Regression
BESO	Bidirectional Evolutionary Structural Optimization
BJ	Binder Jetting
BOM	Bill of Materials
BREP	Boundary Representation
CAD	Computer-aided Design
CAE	Computer-aided Engineering
CAM	Computer-aided Manufacturing
CART	Classification and Regression Tree
CFD	Computational Fluid Dynamics
cGAN	Conditional Generative Adversarial Network
CNC	Computer Numerical Control
CNN	Convolutional Neural Network
DED	Directed Energy Deposition
DFA	Design for Assembly
DfAM	Design for Additive Manufacturing
DFM/DFMA	Design for Manufacturing/Design for Manufacturing and Assembly
DfMAM	Design for Multi-material Additive Manufacturing
DFX	Design for X
DIW	Direct Ink Writing
DNN	Deep Neural Network
DOE	Design of Experiments
EBM	Electron Beam Melting

FDM	Fused Deposition Modeling
FEA	Finite Element Analysis
FEM	Finite Element Method
FMEA	Failure Mode and Effects Analysis
FS	Functional Surface
FV	Functional Volume
GAN	Generative Adversarial Network
G-CNN	Graph-informed Convolutional Neural Network
IDS	Initial Design Space
LCA	Life Cycle Assessment
LPBF	Laser Powder Bed Fusion
MBID	Multifunctional Bio-inspired Design
MDN	Mixture Density Network
ML	Machine Learning
MMAM	Multi-material Additive Manufacturing
MRI	Magnetic Resonance Imaging
PBF	Powder Bed Fusion
PC	Part Consolidation
PLA	Polylactic Acid
RBF	Radial Basis Function
RL	Reinforcement Learning
RMSE	Root Mean Square Error
SBF	Support-build Feature
SIMP	Solid Isotropic Material with Penalization
SLA	Stereolithography
SLM	Selective Laser Melting
SLS	Selective Laser Sintering
STL	Stereolithography
SVM	Support Vector Machines
TO	Topology Optimization
TPU	Thermoplastic Polyurethane
UCB	Upper Confidence Bound
UV	Ultraviolet
VAE	Variational Autoencoder
VP	Vat Photopolymerization

1

Introduction

This chapter serves as the foundational introduction to the book *Design for Additive Manufacturing – Fundamental Principles and Advanced Strategies*. It begins by providing an overview of key additive manufacturing (AM) technologies to establish a common technical understanding. It then transitions into a discussion of design for additive manufacturing (DfAM), outlining its conceptual basis, current practices, and emerging challenges. The aim is to situate DfAM within the broader context of engineering design and motivate the specialized topics addressed in subsequent chapters, including design methodologies, optimization, lattice structures, digital materials, and data-driven approaches.

1.1 Additive Manufacturing

The fundamental principle of AM lies in building three-dimensional (3D) geometries by adding material, typically in a layer by layer fashion [1]. Despite variations in processes and materials, AM techniques share several core features: the use of a computer to store and process geometric information, a digital model to guide the build path, and a deposition system that processes material in points, lines, or areas to fabricate parts.

According to the ISO/ASTM 52900 standard [2], AM is defined as a "process of joining materials to make parts from 3D model data, usually layer upon layer, as opposed to subtractive manufacturing and formative manufacturing methodologies." AM processes enable the fabrication of complex geometries with minimal material waste and often require little to no post-processing, thus offering significant advantages over traditional subtractive and formative methods.

ISO/ASTM 52900 categorizes AM processes into seven groups:

- **Binder Jetting (BJ)**
- **Directed Energy Deposition (DED)**
- **Material Extrusion (ME)**
- **Material Jetting (MJ)**
- **Powder Bed Fusion (PBF)**
- **Sheet Lamination (SL)**
- **Vat Photopolymerization (VP)**

Each category has unique capabilities, material compatibilities, and application domains. Material type critically affects the final part's geometry, mechanical performance, durability, and cost, thereby influencing the potential applications of AM. Materials are typically categorized into solids, powders, and liquids, spanning polymers, metals, ceramics, and composites [3]. Although the material choices available for AM are still limited compared to traditional manufacturing, the rapid advancement of AM technologies and materials processing methods suggests a future with broader material diversity.

1.1.1 Laser-based Powder Bed Fusion

Among AM technologies, the laser-based powder bed fusion (LPBF) process is a leading method for metal parts production. In LPBF, a laser beam serves as the energy source, selectively melting and fusing layers of fine metal powder according to a computer-generated pattern. After each layer is completed, a fresh layer of powder is spread, and the process repeats until the entire 3D part is built.

A schematic of the LPBF process is shown in Figure 1.1. The essence of LPBF lies in controlling the thermal interaction between the intense laser beam and the low-conductivity powder bed to achieve precise and localized fusion [4]. The fusion mechanisms in LPBF can be categorized as solid-state sintering, chemically induced binding, liquid-phase sintering (LPS), and full melting. For metallic materials, LPS and full melting are the primary modes.

LPBF offers a high level of detail resolution, a broad selection of metallic powders, and the ability to fabricate complex geometries unachievable by traditional means. Applications span medical implants, aerospace components, automotive parts, tooling, and energy sector devices. Compared to conventional manufacturing, LPBF is especially advantageous for prototyping and small-batch production of intricate parts, thanks to reduced tooling needs, shorter lead times, and improved cost efficiency.

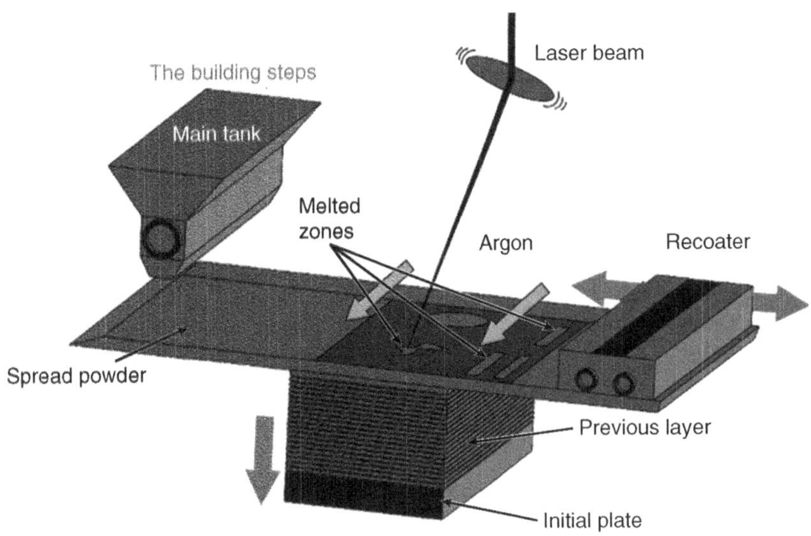

Figure 1.1 Schematic of the LPBF process.

Research and industrial investment in LPBF have grown substantially over the past two decades. This technology is now considered relatively mature, supported by a growing portfolio of printable materials such as AlSi10Mg (aluminum alloy), Ti-6Al-4V (titanium alloy), stainless steel 17-4PH, maraging steel MS1, and Inconel alloys. LPBF is increasingly transitioning from purely prototyping to functional end-use parts manufacturing, marking a pivotal moment in its adoption lifecycle [5].

1.1.2 Material Extrusion

ME is a type of AM process in which a feedstock material (typically a thermoplastic polymer) is forced through a heated nozzle and selectively deposited layer by layer to create a 3D object. The most well-known example is fused deposition modeling (FDM), which is also known as fused filament fabrication. Figure 1.2 demonstrates the schematic of the FDM process [6]. Filaments are used in the printing process. Potential variants of the ME process include different extrusion processes such as the hot extrusion of rods, cold extrusion of slurries, and hot extrusion of pellets. A schematic of the FDM process is illustrated in Figure 1.2.

ME processes are generally slower and less precise than LPBF but offer unmatched accessibility and affordability, making them popular for desktop 3D printers and early-stage prototyping. ME supports a wide range of materials, including PLA, ABS, PETG, nylon, and high-performance thermoplastics like PEEK [7]. It also extends to ceramic slurries, green-state metals, food-grade pastes, and biomaterials. Despite lower resolution compared to other AM technologies, ME remains integral for rapid prototyping, iterative design, and educational applications due to its low barrier to entry.

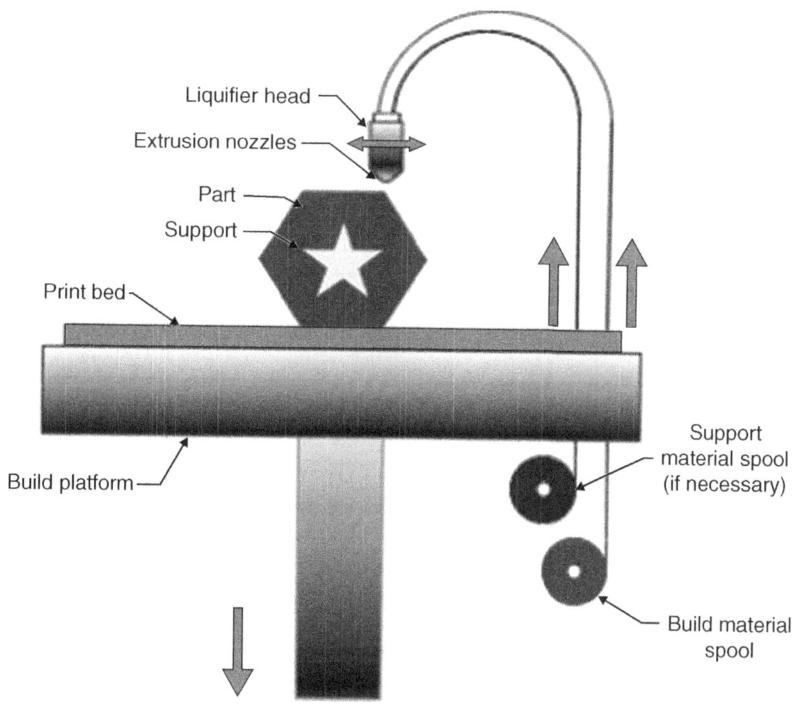

Figure 1.2 Schematic of the FDM process.

1.1.3 Other Additive Manufacturing Processes

1.1.3.1 Binder Jetting (BJ)

In BJ, a liquid binding agent is selectively deposited onto a powder bed to bond powder particles together. After printing, the "green" part usually requires sintering to densify the structure. BJ is used for metals, ceramics, and sand-based molds [8]. Figure 1.3 shows the schematic of BJ process [9].

1.1.3.2 Directed Energy Deposition (DED)

DED utilizes focused thermal energy (laser, electron beam, or plasma arc) to melt and deposit material simultaneously. The process can repair existing components or build new structures, particularly with metals. Different subtypes of DED process exist, and they can be categorized either by their energy source or the material format (powder or wire) [10]. Figure 1.4 shows the schematic of (a) powder DED and (b) wire DED processes.

1.1.3.3 Material Jetting (MJ)

In MJ, droplets of photopolymer or wax are deposited and cured using ultraviolet light [9]. MJ offers high-resolution, multi-material, and full-color printing but is typically limited to prototypes and models. Figure 1.5 shows the schematic of MJ process.

1.1.3.4 Sheet Lamination (SL)

SL involves bonding sheets of material (paper, metal, or polymer) layer by layer using adhesives or ultrasonic welding. Parts are then cut to shape using lasers or mechanical means. It is suitable for low-cost prototypes and large parts. Figure 1.6 shows the schematic of SL process [9].

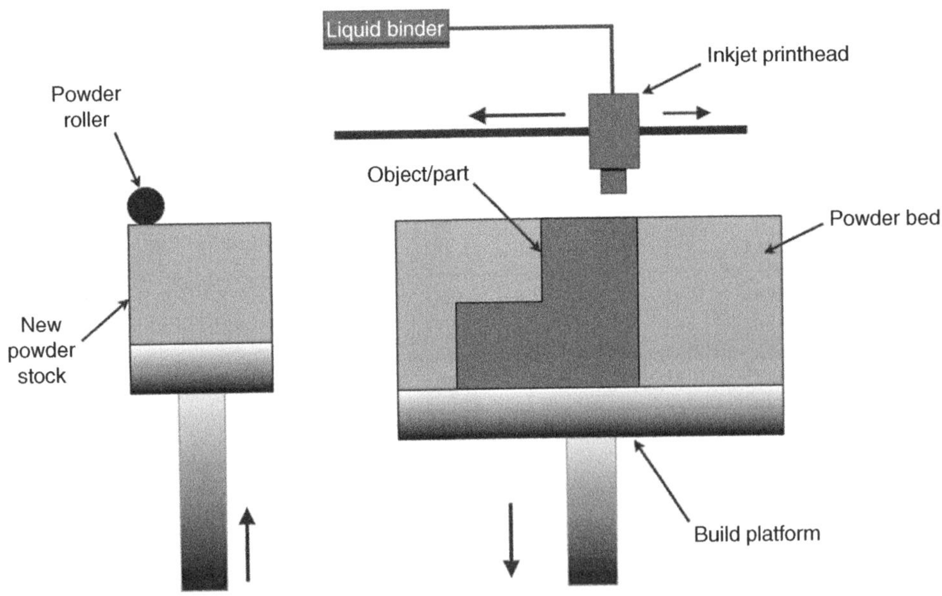

Figure 1.3 Schematic of BJ process.

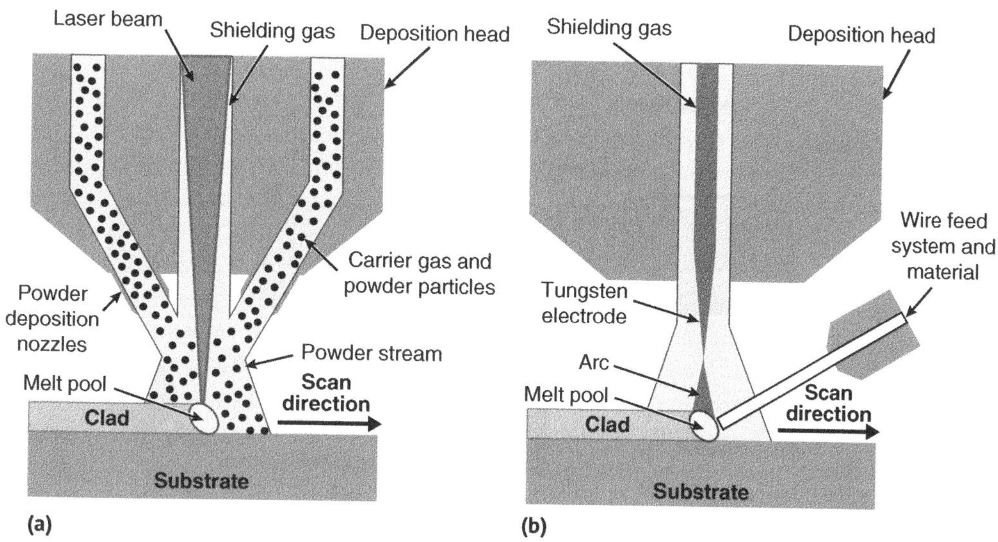

Figure 1.4 (a) Powder DED process and (b) wire DED process.

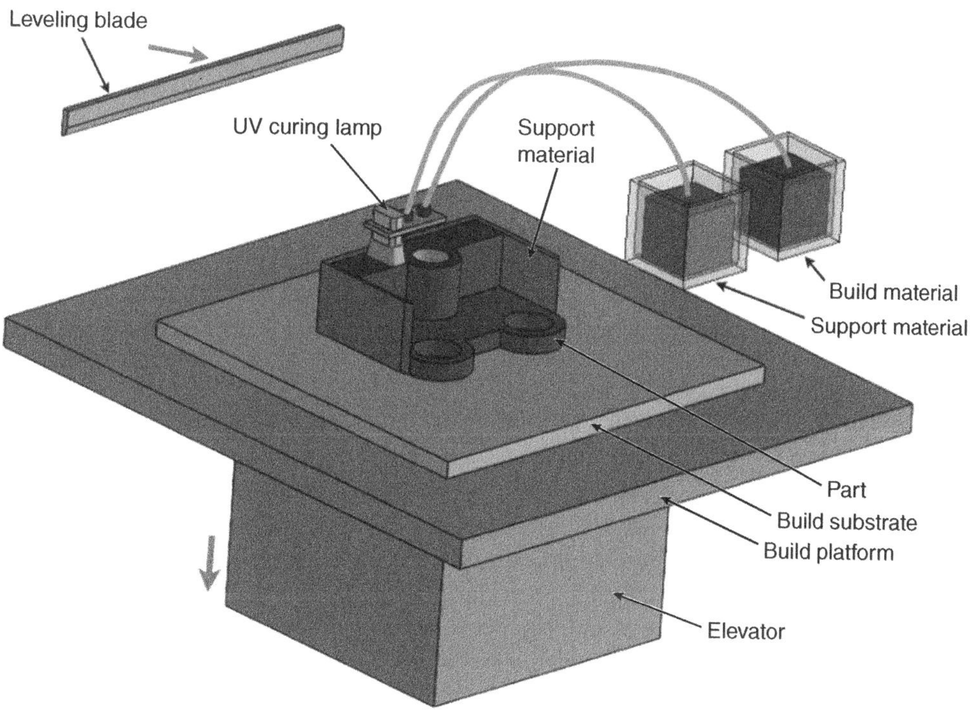

Figure 1.5 Schematic of MJ process.

- Machine Learning-accelerated TO, where models like CNNs and GANs predict optimized structures from intermediate states, reducing iteration times.
- Reliability-based TO, which considers uncertainty and robustness in demanding applications.

TO serves not only as a method for part optimization, but as a core enabler of part consolidation, multi-material integration, and certification-ready design—all of which are central to the DfAM agenda.

5. **Machine Learning and Data-driven Design**
   Beyond TO, ML is revolutionizing DfAM in other areas: feature recommendation in conceptual design, predictive modeling of print outcomes, and generative design for metamaterials. ML is also being used to bridge gaps between simulation fidelity and real-world print outcomes, creating new pathways for closed-loop design workflows [7, 12, 14, 29].
6. **Digital Infrastructure and Workflow Integration**
   Case studies reveal that fragmented digital toolchains (e.g., STL-based workflows, manual input) limit efficiency, increase design cycle times, and introduce knowledge loss. Addressing data compatibility and enhancing interoperability is crucial for advancing DfAM adoption in industry [30, 31].
7. **Multi-scale and Multi-material Design**
   Effective DfAM must account for performance across multiple scales—from microstructures to macro-performance—and increasingly involves multi-material printing, especially in bioengineering and aerospace contexts [1, 3, 32, 33].
8. **Sustainability-driven DfAM**
   DfAM naturally supports sustainability goals through material efficiency, part consolidation, reduced waste and support material as well as designs for recyclability and reusability. Several frameworks evaluate AM sustainability across economic, environmental, and social dimensions. Eco-design strategies such as optimizing overhangs, hollowing structures, or using recycled feedstock are now commonly integrated into DfAM best practices [12, 31, 34].

Various interpretations of DfAM exist throughout this research domain. There is still a need for formalization in these definitions to aid in clearer educational, research, or application frameworks and practices. This book adopts and expands the concept of DfAM and defines it as:

> ***Design for Additive Manufacturing (DfAM) refers to a class of design methods that optimize functional performance and/or other key product life-cycle considerations—such as manufacturability, reliability, and cost—while explicitly accounting for the unique capabilities of additive manufacturing technologies.***

This perspective positions DfAM at the intersection of many traditional design for X (DFX) strategies, such as design for manufacturing, design for maintenance, and design for cost. However, unlike traditional DFX approaches, DfAM uniquely enables designers to embrace design complexity as a resource—rather than treating it as a liability. The imperative for dedicated design research in additive manufacturing stems from several key factors. Firstly, the very design freedoms offered by AM necessitate new methodologies and tools that move beyond the constraints of conventional manufacturing. Traditional design rules often stifle innovation and prevent the realization of AM's unique capabilities. Secondly, the complex interplay between design parameters, AM process variables, and material properties requires in-depth investigation to ensure predictable and reliable outcomes. Research is crucial for developing design guidelines, simulation tools, and optimization techniques tailored specifically for AM. Ultimately, robust design research will pave the way for the widespread adoption of AM in critical applications by ensuring both the manufacturability and the enhanced performance of AM-produced parts.

1.3 Challenges and Future Directions of DfAM

Despite its rapid advancement, DfAM faces several unresolved challenges that must be addressed to unlock its full industrial potential. These challenges span methodology, computation, infrastructure, validation, sustainability, and organizational practices. The following paragraphs discuss these challenges in detail, alongside future research directions that can guide the field toward more mature, integrated, and effective design practices.

1.3.1 Design Knowledge and Methodology Gaps

DfAM is still maturing as a formal design discipline. Much of its practice relies on implicit knowledge or context-dependent heuristics that have not been systematically validated in industrial settings. The terminology surrounding design heuristics, principles, and guidelines is inconsistent across literature, which complicates their integration into structured design workflows. Emerging frameworks such as dual DfAM [35], which seek to balance manufacturability constraints and design opportunities, are promising but remain underdeveloped for early-stage conceptual design. To advance the field, there is a clear need for validated and structured frameworks that encompass the full design process and incorporate manufacturability feedback. Formalized libraries of design rules—particularly for complex geometries such as lattices—would support broader adoption. Additionally, future research should investigate human-in-the-loop environments that combine heuristic guidance with digital verification and feedback.

1.3.2 Toolchain Fragmentation and Digital Infrastructure Deficiencies

A major bottleneck in DfAM is the fragmentation of design tools and digital infrastructure. Many industrial workflows still rely on limited or outdated file formats like STL, which strip away critical metadata such as tolerances, material gradients, and design intent. This leads to inefficient, error-prone manual processes and disconnected iterations between CAD, simulation, and slicing tools. Improving digital interoperability is essential. Future efforts should focus on adopting next-generation data standards, such as AMF or 3MF, and creating platform-level toolchains that preserve semantic design information [7, 14]. Integrated digital environments and digital twins will also play a central role in bridging design, simulation, and production.

1.3.3 Limitations in Topology Optimization and Generative Design

TO has become a cornerstone of DfAM but still suffers from limitations. Traditional TO methods are computationally intensive and often require extensive iteration. Moreover, they frequently produce geometries that violate AM constraints, necessitating costly post-processing or reinterpretation. There is growing demand for TO methods that are explicitly aware of additive constraints such as overhang angles, print anisotropy, and support minimization. In parallel, machine learning offers promising accelerations to TO workflows, allowing for real-time prediction of optimized structures. Future research should also explore robust, multi-physics, and reliability-based TO frameworks that support multifunctional performance and certification.

1.3.4 Machine Learning and Data-driven Design Challenges

Although machine learning is gaining traction in DfAM, its deployment remains limited by data scarcity, generalizability issues, and the opacity of model reasoning. Most ML models are trained on narrow datasets and may not transfer well across domains or geometries. Additionally, black-box models present challenges for interpretability, especially in high-risk applications requiring certification [36]. To overcome these barriers, future work should integrate ML with physics-based models and design rules in hybrid workflows. Efforts should also focus on making ML models more explainable and developing data-sharing protocols that support cross-domain learning.

1.3.5 Modeling, Simulation, and Validation Limitations

Simulation and validation remain a weak link in DfAM toolchains. Multi-scale and multi-material modeling are still underdeveloped for AM's complex architectures. The lack of benchmark datasets and standardized validation protocols makes it difficult to evaluate design tool performance in realistic industrial contexts. Future progress requires robust co-simulation environments that combine structural, thermal, and process modeling, as well as empirical validation through standardized testbeds and industry partnerships.

1.3.6 Material and Process Constraints

DfAM is inherently tied to the constraints and opportunities of the underlying AM process and material system. However, current design tools often ignore or oversimplify these influences. Material properties in AM are frequently anisotropic and process-dependent, which complicates accurate performance prediction. Additionally, the range of usable materials remains limited for many AM methods. Addressing these issues demands more comprehensive material-process-performance databases, improved predictive models, and adaptive design systems that can dynamically respond to material constraints.

1.3.7 Sustainability and Life Cycle Considerations

Although AM can reduce waste through near-net-shape fabrication, DfAM strategies often neglect broader lifecycle and sustainability implications. Energy consumption, material sourcing, and end-of-life design are rarely integrated into current tools. There is a need to embed environmental considerations such as carbon footprint, recyclability, and material circularity into early-stage design decisions. Eco-design features like hollowing parts, minimizing supports, and using recycled feedstock must become standard practice in DfAM strategies.

1.3.8 Human-centric and Organizational Barriers

Finally, organizational and educational challenges remain. Effective DfAM requires interdisciplinary collaboration, yet silos persist between design, engineering, and manufacturing teams. Furthermore, many designers lack the training needed to fully exploit AM's capabilities. Addressing these gaps calls for user-centric design platforms, accessible training resources, and cross-functional team structures. Establishing formal DfAM curricula and certification programs will be key to fostering widespread expertise.

The concepts introduced in this chapter establish the foundation for the chapters that follow. Chapter 2 presents the design process framework and highlights how traditional design phases—from function modeling to detailed design—must adapt for additive manufacturing. Chapter 3 explores methods for early-stage DfAM, including design heuristics, worksheets, and conceptual tools. Chapter 4 addresses strategies for part consolidation and assembly redesign, while Chapter 5 introduces AM-specific constraints and process-aware design rules. Chapters 6 and 7 delve into advanced optimization and structural design approaches, focusing on topology optimization, multifunctional performance, and lattice structures. Chapter 8 introduces digital materials and their fabrication challenges, and Chapter 9 concludes by examining emerging data-driven and bio-inspired design strategies that signal future directions in DfAM research and practice.

References

1 Gao, W., et al., *The status, challenges, and future of additive manufacturing in engineering*. Computer-Aided Design, 2015. **69**: pp. 65–89.

2 ASTM International, *Standard terminology for additive manufacturing — general principles — terminology*, ASTM F52900-21, West Conshohocken, PA: ASTM International, 2021.

3 Bourell, D., et al., *Materials for additive manufacturing*. CIRP Annals - Manufacturing Technology, 2017. **66**(2): pp. 659–681.

4 Sun, S., Brandt., M., and Easton, M., 2 - Powder bed fusion processes: an overview, in *Laser additive manufacturing*. M. Brandt Editor. 2017, Woodhead Publishing. pp. 55–77.

5 Aiza, I., et al., *Effects of build orientation and inclined features on physical, microstructural and mechanical properties of powder bed fusion additively manufactured metallic parts*. Progress in Materials Science, 2025. **147**: p. 101357.

6 Ning, F., et al., *Additive manufacturing of carbon fiber reinforced thermoplastic composites using fused deposition modeling*. Composites Part B: Engineering, 2015. **80**: pp. 369–378.

7 Lopez Taborda, L.L., Maury, H., and Esparragoza, I.E., *Design methodology for fused filament fabrication with failure theory: framework, database, design rule, methodology and study of case*. Rapid Prototyping Journal, 2024. **30**(9): pp. 1803–1821.

8 Saimon, A.I., et al., *Advancing additive manufacturing through deep learning: a comprehensive review of current progress and future challenges*. IISE Transactions, 2025. **58**(1): pp. 33–56.

9 Loughborough University, *Additive manufacturing research group* (AMRG), *the seven categories of additive manufacturing*, Loughborough University, Loughborough, UK, 2025. Available at: https://www.lboro.ac.uk/research/amrg/about/the7categoriesofadditivemanufacturing/ (accessed April 2025).

10 Nuñez, L., et al., *Analysis of surface roughness in metal directed energy deposition*. The International Journal of Advanced Manufacturing Technology, 2024. 132(11–12): pp. 1–20.

11 Zhang, F., et al., *The recent development of vat photopolymerization: a review*. Additive Manufacturing, 2021. **48**: p. 102423.

12 Lopez Taborda, L.L., Maury, H., and Pacheco, J., *Design for additive manufacturing: a comprehensive review of the tendencies and limitations of methodologies*. Rapid Prototyping Journal, 2021. **27**(5): pp. 918–966.

13 Laverne, F., et al., *Assembly based methods to support product innovation in design for additive manufacturing: an exploratory case study*. Journal of Mechanical Design, 2015. **137**(12): p. 121701.

14 Hanks, B., et al., *Mechanical properties of additively manufactured metal lattice structures: data review and design interface*. Additive Manufacturing, 2020. **35**: p. 101301.

15 Khan, N. and Riccio, A., *A systematic review of design for additive manufacturing of aerospace lattice structures: current trends and future directions*. Progress in Aerospace Sciences, 2024. **149**: p. 101021.

16 Tang, Y., et al., *Lattice structure design and optimization with additive manufacturing constraints*. IEEE Transactions on Automation Science and Engineering, 2018. **15**(4): pp. 1546–1562.

17 Tang, Y., Kurtz, A., and Zhao, Y.F., *Bidirectional Evolutionary Structural Optimization (BESO) based design method for lattice structure to be fabricated by additive manufacturing*. Computer-Aided Design, 2015. **69**: pp. 91–101.

18 Yang, S., Tang, Y., and Zhao, Y.F., *A new part consolidation method to embrace the design freedom of additive manufacturing*. Journal of Manufacturing Processes, 2015. **20**: pp. 444–449.

19 Letov, N. and Zhao, Y.F., *A geometric modelling framework to support the design of heterogeneous lattice structures with non-linearly varying geometry*. Journal of Computational Design and Engineering, 2022. **9**(5): pp. 1565–1584.

20 Valjak, F. and Lindwall. A., Review of design heuristics and design principles in design for additive manufacturing. in *Proceedings of the Design Society*, 2021.

21 Yang, S., Page, T., and Zhao, Y.F., *Understanding the role of additive manufacturing knowledge in stimulating design innovation for novice designers*. Journal of Mechanical Design, 2018. **141**(2): p. 021703.

22 Yang, S. and Zhao, Y.F., *Additive manufacturing-enabled design theory and methodology: a critical review*. The International Journal of Advanced Manufacturing Technology, 2015. **80**(1): pp. 327–342.

23 Anh, N.T., Quynh, N.X., and Tung, T.T., *Study on topology optimization design for additive manufacturing*. Engineering, Technology and Applied Science Research, 2024. **14**(3): pp. 14437–14441.

24 Hertlein, N., Vemaganti, K., and Anand, S., *Design optimization of lattice structures under impact loading for additive manufacturing*. Journal of Mechanical Design, 2024. **146**(11): p. 111701.

25 Kongprasert, N. and Nguyen, D.S., *Driving innovation in product design: integrating additive manufacturing and topology optimization techniques*. Journal of Applied Science and Engineering, 2025. **28**(1): pp. 53–60.

26 Lkadi, O., Nassraoui, M., and Bouksour. O., Design for additive manufacturing: opportunities, limits, and methods. in *AIP Conference Proceedings*. 2024.

27 Mohseni, M. and Khodaygan, S., *Design for additive manufacturing of topology-optimized structures based on deep learning and transfer learning*. Rapid Prototyping Journal, 2024. **30**(7): pp. 1411–1433.

28 Wyetzner, S.D.T., et al., *Regenerative topology optimization of fine lattice structures*. 3D Printing and Additive Manufacturing, 2023. **10**(2): pp. 183–196.

29 Wang, Z., et al., *Implications of data-driven product design: from information age towards intelligence age*. Advanced Engineering Informatics, 2022. **54: p**. 101793.

30 Mallalieu, A., et al. The role of digital infrastructure for the industrialisation of design for additive manufacturing. in *Proceedings of the Design Society*, 2022.

31 Tang, Y., Mak, K., and Zhao, Y.F., *A framework to reduce product environmental impact through design optimization for additive manufacturing*. Journal of Cleaner Production, 2016. **137**: pp. 1560–1572.

32 Algunaid, K.M.A. and Liu, J., *Decision support system to select a 3D printing process/machine and material from a large-scale options pool*. International Journal of Advanced Manufacturing Technology, 2022. **121**(11-12): pp. 7643–7659.

33 Liu, Y., et al., *A survey of modeling and optimization methods for multi-scale heterogeneous lattice structures*. Journal of Mechanical Design, 2020. **143**(4): p. 040803.

34 Alfaify, A., et al., *Design for additive manufacturing: a systematic review*. Sustainability (Switzerland), 2020. **12**(19): p 7936.

35 Dash, S., Nordin, A., and Johansson, G., *Dual design for additive manufacturing in engineering design: a systematic literature review*. Rapid Prototyping Journal, 2024. **31**(11): pp. 40–61.

36 Dilberoglu, U.M., et al. *The role of additive manufacturing in the era of industry 4.0*. Procedia Manufacturing, 2017. **11**: pp. 545–554.

2

Engineering Design Process

Engineering design is one of the most important aspects of engineering and is the primary way in which engineers contribute to society [1]. The engineering design process requires a combination of knowledge and skills of humans, including fundamental subject matter expertise as well as problem-solving and creativity. Engineering design process can be divided into five stages [2]:

1. Problem Definition: The design objectives, constraints, functions, and other information are identified during the problem definition stage to establish a list of customer, client, or end-user requirements. The goal of this stage is detailed and precise list of design requirements, design statements, objectives, constraints, and functions obtained in this stage are input to the subsequent conceptual design stage.
2. Conceptual Design: In this stage, the requirements obtained from the previous stage are translated into engineering design specifications used to benchmark potential designs. Designers harness their creativity and intuition gained from their experience in the conceptual design stage to assist ideation process to generate new, appropriate, and improved design solutions [3, 4]. Through this ideation process, multiple design alternatives are generated according to the engineering design specifications. These multiple design concepts are input to the next design stage for modeling, evaluation, and analysis.
3. Preliminary Design: In this stage, the alternate design concepts are modeled by utilizing computer-aided design (CAD) software to express the design concepts for effective visualization. These models are evaluated based on the performance and function requirements of the customer. This process mainly leverages computational tools such as CAD and simulation software for quick and effective analysis of the design concepts. After the analysis, designers use their decision-making skills to select the proper designs suitable to the requirements.
4. Detailed Design: In this stage, from the evaluations in the previous stage, the selected design is modeled in detail using CAD software. Various types of optimizations and improvements are carried out at this stage. The designers gather information (from handbooks, standards, databases, etc.) and rely on their knowledge and experience to decide the detailed geometry, dimensions and tolerances.
5. Design Communication: In this final stage of the engineering design process, the final design with its details is documented and communicated in the form of reports, engineering drawings, manufacturing specifications, and prototypes. The entire design process is also documented since it may prove beneficial for future reference. The design communication process

can also help obtain feedback from customers and stakeholders to modify the design further and provide their final approval.

This chapter summarizes key concepts and actions carried out in the engineering design processes.

2.1 Design Function Analysis

Mechanical design begins with understanding what a product must accomplish—its functions. This process is often referred as design function analysis, which includes functional modeling, functional decomposition, and function reasoning. This section focuses on functional modeling and functional decomposition, which are often carried out iteratively to refine the functional model and the final decomposed sub-functions. Functional modeling is a philosophy where functional information is abstracted from or deployed to physical artifacts through design activities according to existing physical laws and axioms. Some literature suggests that there are mainly three formal functional modeling approaches [5–7]: flow-based methods [8–10]; behavior-involved methods (including function-behavior-state [6], function-behavior-structure [11], and structure-behavior-function (SBF) method [12]); and functional interpretation language [13] and ontology methods [14]. Functional modeling provides an abstract yet direct way to represent the intended operation of a product [15]. It not only defines the purpose of the design in engineering terms but also strategically guides activities like concept generation and physical modeling. For example, developing a function structure helps in product architecture decisions by distinguishing what sub-functions will be handled by which components or subsystems [16]. Take the FSB [11] method as an example, functional modeling can describe a design in three domains: function (the teleological purpose of the object—what it is for), behavior (the object's actions or reactions—what it does, derived from its structure), and structure (the physical components and their relationships—what it consists of). In FBS, function is connected to structure only through behavior; that is, designers determine how a structure's behaviors realize the intended functions. This approach supports function analysis by reminding designers to map the intended function to required behaviors (performance metrics, etc.) before deciding on a concrete structure. It is a conceptual tool that complements functional decomposition.

Functional decomposition is a core method where an overall product function is systematically broken down into sub-functions [16]. Pahl and Beitz's [17] systematic design approach based on VDI 2221 [18] emphasizes defining an overall function and mapping it into a structured hierarchy of sub-functions (often visualized as a function tree or network) that describe what the product must do in abstract terms. This function structure captures the flow of energy, material, and information between sub-functions, ensuring all required operations are identified without prematurely prescribing how to implement them. By keeping the decomposition solution-neutral, designers can explore a wider solution space in later stages [16]. In practice, initial functional models may be refined iteratively throughout functional decomposition process as understanding improves, but the goal is a clear, solution-independent description of the design problem.

Comparative studies of these existing functional modeling methods indicate that the prerequisites of a functional modeling method should include how to define a function, how to represent or model a function, and how to break a function into sub-functions, in consistence with the call-for-action research on establishing a benchmarking standard protocol for functional modeling comparison [19]. Standard function vocabularies were proposed [15], allowing different teams to mean the same thing (i.e., "transport fluid" or "convert energy" are the standard ways of modeling design functions by creating a verb + noun phrase). This improves communication and enables reuse of functional models across projects.

In summary, design function analysis—through techniques like functional modeling and decomposition—establishes a solid foundation of what needs to be done before deciding how to do it. This foundation reduces the risk of missing critical functions and encourages innovative solutions not biased by preconceived embodiments.

2.2 Concept Generation

Once the functions are clear, designers explore conceptual solutions where many possible design concepts are generated. A variety of methods are used to spark creativity and systematically broaden the solution space:

- **Brainstorming and Idea Generation:** Classic brainstorming remains a popular technique for teams to generate a flurry of ideas. Key principles, first articulated by Alex Osborn [20], are to defer judgment and aim for quantity, i.e., "no such thing as a bad idea" in the early phase. By suspending criticism, teams encourage wild ideas and build on each other's thoughts, often yielding surprising concepts. Variations like 6-3-5 brainwriting provide structured brainstorming: six participants each write down three ideas every five minutes, passing their ideas to others in round-robin fashion [21]. In 30 minutes, this method can generate up to 108 raw ideas, leveraging the cumulative creativity of the group. Such ideation techniques, along with mind-mapping and sketching exercises, help break fixation and ensure a divergent exploration of concepts.
- **Morphological Charts:** To systematically explore combinations of sub-solutions, engineers use morphological analysis. A **morphological chart** lists the essential sub-functions identified for the design (from the function analysis) and brainstorms multiple means to achieve each sub-function [22]. By mixing-and-matching one means for each sub-function, a plethora of complete concepts can be formed. This technique provides a structured map of the *solution space*: for each function there may be several solution principles, and the chart captures these alternatives and their possible combinations. By considering different combinations, designers can discover nonobvious concepts (including hybrid solutions) that satisfy all functions. In essence, morphological charts ensure a systematic coverage of possibilities, rather than relying on ad-hoc creativity alone. It's a powerful tool to generate concept variants that might never emerge from unstructured brainstorming.
- **TRIZ (Theory of Inventive Problem Solving):** Originating from Genrich Altshuller's analysis of patents [23], TRIZ offers a knowledge-based approach to concept generation. Altshuller observed recurring patterns in how inventive engineers resolved contradictions and distilled these into 40 inventive principles and a contradiction matrix. TRIZ guides designers to frame their design conflicts (e.g., strength vs. weight, or speed vs. accuracy) and then look up inventive principles (like "Segmentation," "Local Quality," "Avoiding Symmetry," etc.) that have solved similar contradictions in other fields. By applying these abstract principles, designers can generate innovative concepts that overcome trade-offs in unconventional ways.
- **Concept Evaluation and Selection:** After generating a pool of candidate concepts, a crucial step is deciding which concept(s) to pursue. Two complementary approaches are often used:

decision matrices and axiomatic design (AD). A popular industry technique is the Pugh concept selection method (decision-matrix approach) [24]. The design team identifies criteria based on the design requirements (performance, cost, safety, etc.) and then rates each concept against these criteria relative to a reference (baseline) concept. This structured comparison highlights the strengths and weaknesses of each concept. Through iterative improvement (e.g., combining the best features from different concepts—creating hybrids), the Pugh matrix helps converge on an optimal or consensus concept. On the other hand, AD [25] provides a theoretical framework to judge concepts against high-level principles. Suh's two axioms are: 1) maintain the independence of functional requirements and 2) minimize the information content (complexity) of the design. In practice, the Independence Axiom means a concept in which design parameters each affect only one function (uncoupled design) is superior—it will be easier to adjust and more robust. The Information Axiom favors simpler, less complex designs that meet requirements—effectively maximizing probability of success. Designers can use these axioms as selection criteria: a concept that violates functional independence (parameters are tangled) or has excessive complexity is flagged as risky. Thus, AD provides a scientific lens to complement scoring matrices.

2.3 Geometric Modeling

With a concept selected, engineers proceed to preliminary design state in which geometric modeling is the core activity to create detailed shape representations of the product. Modern design heavily relies on CAD software for this task, which offers various modeling approaches:

- **Solid Modeling** is the standard for mechanical design, producing fully defined 3D geometries with mass properties and interference checks. Two main methods are used: constructive solid geometry (CSG), which builds models from primitive shapes via Boolean operations, and boundary representation (B-rep), which defines solids by their enclosing surfaces. Most CAD systems use B-rep internally due to its flexibility. Solid models serve as the single source of truth, enabling design analysis, engineering drawings, and direct manufacturing [26].
- **Surface Modeling** is used for complex, visually driven geometries such as automotive exteriors or consumer products. It focuses on sculpting high-quality surfaces without volume, typically using NURBS. While powerful for form exploration and aesthetics, surface models must often be thickened or merged with solids to ensure manufacturability [26].
- **Parametric and Feature-based Design** drive design with parameters and constraints rather than fixed geometry. Engineers define rules—e.g., a hole centered on a plate—and CAD systems maintain relationships when dimensions change. Feature-based design builds geometry as a sequence of operations (extrudes, fillets, etc.), offering flexible, modifiable models. Parametric CAD (e.g., SolidWorks, Creo, CATIA) supports iteration, optimization, and scripting, making it ideal for refining complex designs over time [27].
- **Digital Mock-ups and Assemblies** integrate multiple components with realistic mating constraints. Engineers verify clearances, motion, and system interactions, enabling early detection of design issues. Assembly modeling supports exploded views, bill of materials (BOM) generation, and top-down design strategies where parts are built in context to ensure alignment and fit.

In summary, geometric modeling transforms design concepts into precise digital blueprints, supporting analysis, communication, and manufacturing readiness.

2.4 Detailed Design

Detailed design is the phase where the chosen concept is refined into a complete, final design ready for production. Here, every aspect of the product is fleshed out to meet functional requirements, comply with standards, and optimize manufacturability. Key activities in detailed design include:

- **Engineering Drawings and Tolerancing:** Even in the age of 3D models, technical drawings remain a crucial deliverable of detailed design. Drawings provide a controlled document that specifies dimensions, tolerances, finishes, and assembly instructions in a human-readable format (often required on shop floors or for quality inspection). Geometric dimensioning and tolerancing (GD&T) [28] is applied to define the allowable variation for each feature in a precise symbolic language. GD&T, standardized by ASME Y14.5 (and ISO GPS standards), communicates how far real parts may deviate from the ideal geometry. By using GD&T symbols for flatness, perpendicularity, concentricity, etc., the designer ensures the part will function and assemble correctly even with inevitable manufacturing variations. Tolerancing are decided as part of the design process: too tight a tolerance drives up cost or may be infeasible; too loose a tolerance might cause malfunctions. Thus, detailed design involves careful trade-offs to specify tolerances that balance performance and manufacturability. The result is a set of drawings or model-based definition (MBD) when CAD is used to establish the geometry of the design and tolerances are annotated, which become part of the contractual specification for manufacturing. These are often checked through design reviews and against drafting standards to ensure clarity. In addition to geometry, drawings and documents will include BOMs, identifying each part or standard component (with material and quantity), and specifications in the form of notes on drawing documents for specific manufacturing requirements. This complete documentation package enables a smooth transition to prototyping and production, with no ambiguity about what is to be made.
- **Material and Process Selection:** Detailed design narrows down material choices for each part and the manufacturing processes to be used. During concept phase, one might assume a generic material (e.g., aluminum alloy), but now the specific grade (e.g., 6061-T6 aluminum) and suitable manufacturing process (casting, extrusion, etc.) must be selected. Material selection is often guided by systematic methods: designers use material property databases and charts (like Ashby chart [29]) to choose materials that meet performance requirements, such as strength, stiffness, density, temperature resistance, etc., while minimizing cost and weight. Accompanying material selection process, design simulations using computer-aided engineering (CAE) tools such as finite element analysis (FEA) might be used to predict stress and deflection in critical areas, leading to minor design modification/refinement (adding ribs or fillets). Optimization algorithms might be run on the detailed CAD model to fine-tune dimensions for weight or performance. Material selection is also closely tied to manufacturing process selection: the geometry and tolerances dictate certain processes (die casting vs. machining vs. additive manufacturing, etc.), and each process in turn limits the materials. Designers often consult manufacturing experts and standards to ensure the chosen process can achieve the required geometry and tolerance. The outcome of this task is that every part has a specified material and a suggested manufacturing route. By locking down materials and processes in detailed design, the engineer ensures the design is not only functional but also ready to produce with known techniques and supply chain.
- **Design for X (DfX):** As part of detailed design optimization, engineers apply various Design for Excellence [30] guidelines, where "X" can be Manufacture, Assembly, Reliability,

Safety, etc. Notably, Design for Manufacturing and Assembly (DFMA) principles are used to simplify the product structure and ease fabrication. DFMA may involve reducing part count (combining functions into one part if possible), choosing features that can be machined or molded easily, and designing parts to orient and assemble in only one way (mistake-proof assembly). For example, if two parts could be joined permanently in the factory, DFMA might suggest integrating them into one casting to eliminate a fastening step—if it doesn't complicate other requirements. These principles, documented by Boothroyd and Dewhurst [31] and others, help cut costs and assembly time. Considerations such as reliability, safety, life cycle, modularity, etc. may start in concept phase, they become concrete in detailed design—for instance, performing a DFMA analysis on the near-final design to identify any unnecessary complexity. Incorporating DfX in detailed design ensures the product is not only functional but also optimized for real-world production, use, and lifecycle concerns.

- **Prototyping and Testing:** Before committing to full production, detailed design is validated through prototypes and tests. Prototyping can range from rapid prototypes (using 3D printing or simple machining) to fully functional initial prototypes built with near-final materials. The goal is to test the design against the requirements to find answers to questions such as does it fit, does it withstand loads, and does it perform as expected? In the detailed design stage, tests are often more specific: environmental tests (thermal, vibration), life cycle tests (fatigue, wear), and user tests for ergonomics. All these verification and validation activities ensure that by the end of detailed design, the product meets specifications and no major issues remain. If problems are discovered, the design is iterated (back to some of the prior steps). Finally, when tests indicate the design is sound, the detailed design phase concludes with a "design freeze." The outcome is a fully defined product: 3D models, drawings, material specs, process plans, and validated performance, ready for pilot production and launch. These documentations are part of the design communication step, which is not further discussed in this chapter.

2.5 Computational Design

In recent decades, computational design techniques [32] have emerged as powerful extensions to traditional methods, leveraging algorithms and computer intelligence to aid or even automate aspects of the design process. These approaches are still rooted in classical design principles but provide a generative and data-driven angle to engineering design. Key facets of computational design include:

- **Generative Design and Optimization:** Generative design refers to using algorithms (often AI-driven or optimization algorithms) to automatically produce design alternatives based on defined goals and constraints [33]. Instead of manually tweaking a design, the engineer specifies the requirements—e.g., a component must withstand certain loads, fit within a space, use a given material, and maybe minimize weight—and the generative design algorithms explore numerous shapes that satisfy those criteria. Modern generative design tools integrate with CAD and use techniques like topology optimization, evolutionary algorithms, and even machine learning to evolve designs. For example, topology optimization starts with a design space and removes inefficient material, yielding organic-looking structures that meet performance targets with minimal weight. Generative design in a broader sense can incorporate manufacturing constraints (like only produce geometries that can be 3D-printed or CNC machined). Generative design methods greatly amplify the exploration of design space, often finding nonintuitive solutions that a human might overlook. In practice, generative design is used in aerospace, automotive, and other fields to push the limits of performance (e.g., Airbus used it for partition designs, resulting in a nature-inspired form). It is important to note that generative design still requires a well-formulated problem definition by the engineer—the quality of results depends on having appropriate constraints and objectives.

- **Algorithmic Design:** While parametric CAD was discussed earlier to encode design intent, algorithmic design takes it a step further by using scripting and programming to create complex geometry or to automate design workflows [34]. Many CAD platforms provide APIs or visual scripting tools (like Rhino's Grasshopper, Siemens NX Open, etc.) that let designers define algorithms that generate geometry. This is sometimes called design scripting or procedural modeling. For instance, instead of manually modeling each variation of a gear, an engineer could write a script that takes number of teeth, module, and pressure angle as inputs and automatically builds the 3D model of the gear. Algorithmic design is powerful for creating families of parts, lattices or repeating patterns, and any design governed by mathematical rules. It is also used in architecture and product design for freeform patterns that would be tedious to model by hand. In mechanical engineering, a designer can use it to optimize a structure by writing an algorithm to vary a truss geometry and call an FEA solver in the loop to converge on an optimal topology.
- **Artificial Intelligence (AI) in Design:** AI is increasingly influencing engineering design methodology. A recent survey [1] observed seven AI-based methods used in different design stages. They are: 1) agent-based modeling, 2) deep learning, 3) machine learning, 4) network/graph theory, 5) NLP, 6) optimization algorithms, and 7) probabilistic methods. Among them, deep learning is the most used AI-based method across the design stages, followed by machine learning and NLP. Most AI-based methods used in the design process focus on conceptual and preliminary design stages [35], where generative design-based deep learning and supervised machine learning methods are used for design generation and evaluation. One reason for this situation is the attractive promise of AI algorithms driven by data serve as assistants for designers in quick and effective, effort-intensive, iterative design tasks such as ideating, generating, and evaluating design concepts and models. Traditionally, these stages rely on the designer's knowledge, experience, creativity, and decision-making skills to iteratively process, generate, and evaluate multiple design concepts to achieve a creative and innovative solution. Additionally, there are more open repositories providing design data for AI-based analysis and design generation at conceptual and preliminary design stages. However, despite recent advances in AI, their capabilities are far from accurately replicating a designer's creativity and heuristics. In these AI-based research, the AI algorithms mainly support designers by automating or analyzing certain design tasks within these stages.
- **Design Automation and Digital Workflow:** Computational design also implies a seamless digital thread through the design process. CAD models can be linked to analysis (CAE) and optimization tools, and further to CAM (manufacturing), creating an automated pipeline. For example, in an idealized situation, a computational design environment might allow simple actions to update a CAD model, rerun FEA and fatigue life calculations, update cost estimates, and then suggest design changes if certain criteria are not met. However, the ideal situation of an automated digital design flow with these tools integrated is extremely challenging. Specialized scripts and additional software tools are often needed to be developed for product type.

In conclusion, computational design methods enrich the common engineering design practices. They do not replace the fundamentals of understanding functions, brainstorming concepts, or applying sound engineering judgment; rather, they build on those foundations to handle complexity and explore options at a speed and scale impossible to achieve manually. The trajectory of computational design is moving toward more integration and autonomy, but the role of the engineer remains pivotal in guiding the algorithms, setting the objectives, and making final decisions that ensure the design is successful in the real world.

2.6 Bridging the Gap: From Traditional Design to Design for Additive Manufacturing

The design paradigms underpinning traditional mechanical engineering and those required for additive manufacturing (AM) differ in both scope and intention. While conventional methods prioritize design simplification, part standardization, and geometric manufacturability for formative and subtractive processes, design for additive manufacturing (DfAM) embraces geometric complexity, functional integration, and—increasingly—material complexity. Bridging this conceptual and methodological gap is essential for unlocking the full potential of AM in modern design practice. The following is high-level discussion of the gap between traditional engineering design and DfAM. More detailed discussion will be carried out in the subsequent chapters.

1. From Geometric Restriction to Complexity-enabled Innovation
   Traditional design practices often seek to minimize complexity due to cost and manufacturability constraints. In contrast, AM enables the fabrication of intricate geometries such as internal channels, graded lattices, and organic topologies with little penalty. This shift opens new possibilities but also introduces the need for new tools and rules to guide this expanded design space—an approach outlined in Chapters 3 and 4.
2. Functional Integration Through Part Consolidation
   Traditional assembly-based design discourages component integration due to concerns over tooling, material changes, and assembly flexibility. AM overturns this logic: consolidation reduces part count, minimizes interfaces, and can improve performance through integrated load paths and internal features. Chapter 4 details how to identify candidates for consolidation and fuse functional entities into manufacturable AM parts.
3. Process-aware and Knowledge-driven Conceptualization
   Design conceptualization for AM requires embedding process knowledge from the outset. This includes using tools like AM-specific heuristics, function-based design worksheets, and predictive machine learning models to assess print feasibility or AM suitability. Chapter 3 provides a comprehensive overview of such frameworks, which allow designers to move beyond intuition-driven concept development.
4. Incompatibility of Traditional DfM/DfA with AM-specific Constraints
   Design for Manufacturing and Assembly (DfM/DfA) focuses on draft angles, tool access, and standard fittings, none of which translate directly to AM. Instead, AM presents its own set of constraints—overhang angles, support material removability, thermal distortion, and anisotropic strength—that must be anticipated during the design process. Chapter 5 introduces these process-specific constraints, while Chapters 6 and 7 provide design guides tailored to polymeric and metal AM processes.
5. From Single-material Objects to Digitally Designed Materials
   The transition from designing structures made of materials to designing materials themselves represents a profound evolution. Chapter 8 introduces digital materials design, including

voxel-level control (e.g., material jetting), functionally graded multi-material structures (e.g., FDM, SLA), and fiber-reinforced composites with tailored anisotropy. These approaches require new workflows that integrate multi-physics simulation, optimization, and even machine learning-based property prediction.

6. Toward a Layered DfAM Mindset
Rather than replacing traditional methods, DfAM demands a layered mindset. Structural geometry, functional integration, manufacturing constraints, and material heterogeneity must all be considered concurrently. The subsequent chapters offer this progression: from abstract concept development (Chapter 3) to geometric integration (Chapter 4), process-aware constraint handling (Chapters 5–7), and ultimately, to digitally controlled material distributions (Chapter 8).

In summary, this book moves beyond the limitations of traditional design frameworks, equipping the reader with a pathway into DfAM that starts with familiar engineering principles and gradually builds the competencies needed to leverage AM's transformative capabilities—not only for shaping parts, but for designing multifunctional, material-rich, and performance-optimized systems.

References

1 Khanolkar, P.M., Vrolijk, A., and Olechowski, A., *Mapping artificial intelligence-based methods to engineering design stages: a focused literature review*. Artificial Intelligence for Engineering Design, Analysis and Manufacturing, 2023. **37**: p. e25.
2 Dym, C.L., *Engineering design: a project-based introduction*. 2013: John Wiley & Sons.
3 Sarkar, P. and Chakrabarti, A., *Ideas generated in conceptual design and their effects on creativity*. Research in Engineering Design, 2014. **25**(3): pp. 185–201.
4 Chakrabarti, A. and Blessing, L.T.M. (Eds.), *An anthology of theories and models of design: philosophy, approaches and empirical explorations*, Springer, London, UK, 2014.
5 Erden, M.S., et al., *A review of function modeling: approaches and applications*. Artificial Intelligence for Engineering Design, Analysis and Manufacturing, 2008. **22**(2): pp. 147–169.
6 Umeda, Y. and Tomiyama, T., *Functional reasoning in design*. IEEE Expert-Intelligent Systems and their Applications, 1997. **12**(2): pp. 42–48.
7 Van Eck, D., McAdams, D.A., and Vermaas. P.E., Functional decomposition in engineering: a survey. in *2007 Proceedings of the ASME International Design Engineering Technical Conferences and Computers and Information in Engineering Conference, DETC2007*. 2008.
8 Chakrabarti, A. and Bligh, T.P., *A scheme for functional reasoning in conceptual design*. Design Studies, 2001. **22**(6): pp. 493–517.
9 Lind, M., *Modeling goals and functions of complex industrial plants*. Applied Artificial Intelligence, 1994. **8**(2): pp. 259–283.
10 Otto, K. N. and Wood, K. L., *Product design: techniques in reverse engineering and new Product Development*, Prentice Hall, Upper Saddle River, NJ, USA, 2001.
11 John, S.G., *Design prototypes: a knowledge representation schema for design*. AI Magazine, 1990. **11**(4): pp. 26–36.
12 Goel, A.K., Model revision: a theory of incremental model learning, in *Machine Learning Proceedings 1991*, L.A. Birnbaum and G.C. Collins, Editors. 1991, Morgan Kaufmann: San Francisco (CA). pp. 605–609.

13 Bell, J., Snooke, N. and Price, C., *A language for functional interpretation of model based simulation*. Advanced Engineering Informatics, 2007. **21**(4): pp. 398–409.
14 Kitamura, Y., et al., *Deployment of an ontological framework of functional design knowledge*. Advanced Engineering Informatics, 2004. **18**(2): pp. 115–127.
15 Hirtz, J., et al., *A functional basis for engineering design: reconciling and evolving previous efforts*. Research in Engineering Design, 2002. **13**(2): pp. 65–82.
16 Motte, D., A review of the fundamentals of the systematic engineering design process models. in *The 10th International Design Conference*. 2008. Dubrovnik, Croatia.
17 Pahl, G., et al., *Engineering design: a systematic approach*. 2007, London: Springer.
18 VDI-Richtlinie, *Methodik zum Entwickeln und Konstruieren technischer Systeme und Produkte*, in *DIVerlag*. 1993.
19 Summers, J., Eckert, C., and Goel, A., Function in engineering: benchmarking representations and models. in *Proceedings of the International Conference on Engineering Design*, ICED, 2013. **2**.
20 Osborn, A.F., *Applied imagination: principles and procedures of creative problem solving*. 1953, New York: Charles Scribner's Sons.
21 Rhorbach, B., *Creative nach Regeln: Methode 635, eine neue Technik zum Losen von Problemen*. Absatzwirtschaft, 1969. **12**.
22 Smith, G., Morphological charts: a systematic exploration of qualitative design space, in *Mechanical Engineering*. 2007. Clemson University.
23 Genrich, A., *The innovation algorithm: TRIZ, Systematic innovation and technical creativity*. 1999, Technical Innovation Center, Inc.: Worcester, MA.
24 Pugh, S, Concept selection: a method that works. in *Proceedings of International Conference on Engineering Design*, 1981. Zürich.
25 Nam, P.S., *Axiomatic design: advances and applications*. 2001, University Press: New York Oxford.
26 Kamrani, A. K., An overview of computer-aided design, in *Methods in product design: new strategies in reengineering*, A. K. Kamrani and S. A. Nasr (Eds.), Springer, Boston, MA, USA, 2006, pp. 1–18.
27 Zhang, T.T., et al., *A review of parametric approaches specific to aerodynamic design process*. Acta Astronautica, 2018. **145**: pp. 319–331.
28 ASME, *Geometric dimensioning and tolerancing (GD&T): ASME Y14.5-2018*, The American Society of Mechanical Engineers, New York, NY, USA, 2018.
29 Ashby, M.F., *Materials selection in mechanical design*. Metallurgia Italiana, 1994. **86**(10): pp. 475–480.
30 Eastman, C.M., *Design for X: concurrent engineering imperatives*. 2012, Springer Science & Business Media.
31 Boothroyd, G. and P. Dewhurst., *Product design for assembly*. 1987, Boothroyd Dewhurst Incorporated.
32 Gu, N. and Xiangyu, W., *Computational design methods and technologies: applications in CAD, CAM and CAE education*. 2012, Igi Global.
33 Wu, J., Qian, X., and Wang, M.Y., *Advances in generative design*. Computer-Aided Design, 2019. **116**(102733): p. 10.1016.
34 Terzidis, K. Algorithmic design: a paradigm shift in architecture. in *Architecture in the Network Society [22nd eCAADe Conference Proceedings/ISBN 0-9541183-2-4]*. 2004, Copenhagen (Denmark).
35 Luo, J., *Data-driven innovation: What is it?* IEEE Transactions on Engineering Management, 2022. **70**(2): pp. 784–790.

3

Design Conceptualization with Additive Manufacturing

Additive manufacturing (AM) offers tremendous design freedoms, such as complicated geometry, multi-material, embedded components, etc., which was difficult to fabricate with traditional manufacturing processes if not possible. This chapter highlights the tools that are available for supporting design conceptualization with considerations of these new AM-enabled design freedoms and constraints.

3.1 Introduction

AM has been well recognized as an important contributor to design innovation in form, materials, and functionality, which is largely due to its unique layer-by-layer working principle [1, 2]. As such, finding an effective way of taking advantage of AM in the conceptual design stage of product development has received lots of attention. The knowledge of what AM can do and what it cannot regarding design is termed as AM knowledge (AMK) [3]. There are two main challenges of embedding AMK in a design conceptualization process. The first challenge is the extraction and categorization of AMK. Typical AMK categories, for example, include lattice structures, part consolidation, support structure, etc. The other challenge is the delivery of AMK in the design conceptualization process. The AMK is commonly represented through various formats, including design heuristics [4], general design process [5], computational tools [6], and feature repositories [7]. These are typically accompanied by illustrative figures, physical prototypes, computer-aided design (CAD) models, descriptive documentation, or a combination of these elements.

Early studies on AMK categorization encompass a broad spectrum of AM aspects, including process knowledge (e.g., operating principles of various AM techniques), machine characteristics (e.g., build volume, speed, and resolution), product design considerations (e.g., AM capabilities, constraints, optimization methods, design workflows, and geometry modeling), as well as supply chain configurations and business models. Dinar and Rosen [8] proposed formal way of formalizing AMK by using web ontology language. Their work focused on the relationship between prismatic design features and process parameters. An extension of this work includes manufacturing knowledge, business model, and design ontology. However, these works mainly concentrated on the restrictive side of AM to secure manufacturability of a part. In contrast, another stream of AMK categorization is more confined to the opportunistic aspect of AM, by which designers are encouraged to take the new design freedoms of AM into consideration to generate more radical solutions.

For instance, Maidin and colleagues [7] categorized 113 AM design features into four groups: user fit requirements, part consolidation, functionality enhancement, and aesthetic improvement. However, assimilating such a large volume of new information can be overwhelming for novice designers, potentially hindering effective application in practice. The effectiveness of imparting AMK in design studies is also encouraging. Yang et al. [3] designed a booklet of AMK that covers several AM potentials. The comparison experiments showed evidence that with the given AMK booklet, novice designers tend to generate more innovative design ideas than the control groups. Other studies [9–11] also revealed that a proper timing of delivering AMK is critical for increasing creativity and innovation in the concept generation stage. In general, opportunistic AMK proves highly valuable during the early stages of design, whereas restrictive knowledge offers limited utility; however, as the design process advances, this relationship reverses, with restrictive AMK becoming more useful and opportunistic knowledge less so [11]. Having AMK education for novice designers does not mean no pitfalls. Designers who are exposed AM potentials produce less feasible design concepts compared to those who only receive training on design for conventional manufacturing [12].

AMK modeling, categorization, and delivery are of high significance to design for AM education. Research on these topics is still ongoing. This chapter will emphasize on some key tools that have been reported on AMK modeling and delivery. These tools will help novice designers regarding early decisions of whether parts are suitable for AM and what types of AM design freedoms could potentially be applied. This early decision will direct the downstream design activities such as part consolidation and topology optimization.

The chapter will start with the design for additive manufacturing (DfAM) worksheet that focuses on a quick assessment of whether a given design is feasible for AM or not. Then, a design heuristic-based approach that aims to assist designers with AMK in the form of cards is introduced. An integrated DfAM method that imparts AMK in a general design process is covered in Section 3.4. A new trend of delivering AMK through automated methods is discussed in Section 3.5. Lastly, a bio-inspired method to support multifunctional design is presented in Section 3.6.

3.2 DfAM Worksheet

The DfAM worksheet is developed by the REID lab at Purdue University [13]. The DFAM worksheet serves a goal of reducing printing failure, improving understanding of limitations of AM, and suggesting redesign directions. The worksheet can be used in the conceptual stage of a new design or redesign when the CAD model is ready. As shown in Table 3.1, the worksheet has considered eight features of AM design, including complexity, functionality, material removal, unsupported features, thin features, stress concentration, tolerances, and geometric exactness. Under each category, 5-scale or 3-scale options are presented. For example, in complexity category, the complexity of a part is described by the difficulty of machining such as the need of repositioning or internal structures. The increase in complexity will result in a lower score in this category, which indicates a higher chance of successful print with AM. Repeat the evaluation in each category and then add the score. If the total score of evaluation for a given design is S, then the suggested direction of improvement will be:

1. *If* $8 \leq S \leq 15$, higher likelihood of success. The part can be directly printed with no modification.
2. *If* $16 \leq S \leq 23$, moderate likelihood of success. The part needs further detailed evaluation.
3. *If* $24 \leq S \leq 32$, consider redesign. The part is likely to fail the print.
4. *If* $33 \leq S \leq 40$, needs redesign. The part will fail the print.

The worksheet can help novice designers to have a rough estimation of the design solution in a structured way. To better assist the use of this worksheet, an example of a water bottle clamp is shown in Figure 3.1. In the complexity category, the clamp does not have complex curvatures but requires repositioning for milling process. In the functionality category, the ring of the clamp will have frequent contact with water bottle such that it must endure over 100 cycles. In material removal category, there are no internal cavities. In unsupported feature category, the clamp has some unsupported features. In thin feature category, the clamp is thick enough (>3 mm). In the stress concentration category, the connection between the ring and base has no fillet. In tolerance category, the ring of the clamp has openings for adjustment so that the tolerance is less critical. In the geometric exactness category, the clamp has medium-size flat surfaces that require good surface quality to mate other parts. Based on the above reasoning, the overall score will be 27, which means redesign will be required.

Table 3.1 The DfAM worksheet categories.

Complexity	Functionality	Material removal	Unsupported features	Sum across Rows	Totals
The part is the same shape as common stock materials, or is completely 2D	Mating surfaces are bearing surfaces, or are expected to endure for 1000+ of cycle	The part is smaller than the same size as the required support structure	There are long, unsupported features	X5	
The part is mostly 2D and can be made in a mill or lathe without repositioning it in the clamp	Mating surfaces move significantly, experience large forces, or must endure 100–1000 cycles	There are small gaps that will require support structures	There are short, unsupported features	X4	
The part can be made in a mill or lathe, but only after repositioning it in the clamp at least once	Mating surfaces move somewhat, experience moderate forces, or are expected to last 10–100 cycles	Internal cavities, channels, or holes do not have openings for removing materials	Overhang features have a slopped support	X3	
The part curvature is complex (splines or arcs) for a machining operation such as a mill or lathe	Mating surfaces will move minimally, experience low forces, or are intended to endure 2–10 cycles	Material can be easily removed from internal cavities, channels, or holes	Overhanging features have a minimum of 45-degree support	X2	
There are interior features or surface curvature is too complex to be machined	Surfaces are purely nonfunctional or experience virtually no cycle	There are no internal cavities, channels, or holes	Part is oriented so there are no overhanging features	X1	

(Continued)

Table 3.1 (Continued)

Complexity	Functionality	Material removal	Unsupported features	Sum across Rows	Totals
Thin Features	**Stress concentration**	**Tolerances**	**Geometric exactness**	+	
Some walls are less than 1.5 mm thick	Interior corners have no chamfer, fillet, or rib	Hole or length dimensions are nominal	The part has large, flat surfaces or has a form that is important to be exact	X5	
Walls are between 1 mm and 3 mm thick	Interior corners have chamfers, fillets, and/or ribs	Hole or length tolerances are adjusted for shrinkage or fit	The part has medium-sized, flat surfaces, or forms that should be closed to exact	X3	
Walls are more than 3 mm thick	Interior corners have generous chamfers, fillets, and/or ribs	Hole and length tolerances are considered or are not important	The part has small or no flat surfaces, or forms that need to be exact	X1	
Overall					

Source: Adapted from [13].

Ningbo Luoting Trade Co., Ltd.

(a)

Complexity	Functionality	Material removal	Unsupported features	Sum across rows	Totals
				x5	0
	■		■	x4	8
■				x3	6
				x2	0
		■		x1	1
Thin features	Stress concentration	Tolerances	Geometric exactness	+	
	■			x5	5
		■	■	x3	6
■				x1	1
				Overall	27

(b)

Figure 3.1 AM feasibility evaluation of a clamp using the DfAM worksheet. (a) Clamp and (b) the scoring table.

3.3 Design Heuristics for AM

A design heuristic is a context-dependent directive understanding, which could provide designers with the direction of reaching a satisfactory but not essentially optimal solution [14]. The goal of creating design heuristics for AM is to form a pool of knowledge representations that can provide inspirations for designers in the early design stage. By studying 275 artifacts that are fabricated with AM, Blosch–Paidosh and Shea [4] developed a set of design heuristics that are classified into eight thematic categories: part consolidation, customization, convey information, materials, embed/enclose, lightweight, material distribution, and reconfiguration. As summarized in Figure 3.2, each category can further be broken down into subgroups. For example, part consolidation category includes four subgroups considering the different objectives including better functional performance, reduced assembly

Tier I	Tier II
Part consolidation	Consolidate parts for better functional performance
	Consolidate parts to reduce assembly time
	Consolidate parts to increase robustness
	Consolidate parts to achieve multiple functions
Customization	Customize geometry to use case
	Customize user interface to use case
	Customize artifact with decoration
Convey information	Convey information with color
	Convey information with geometry
	Convey information with haptics
	Convey information with light
Materials	Use biocompatible material
	Use biodegradable material
	Use metamaterial to achieve recyclability
	Use metamaterial to achieve desired material properties
	Use variable metamaterial to achieve variable material properties
	Use multiple materials to achieve functionality
	Use multiple materials to achieve variable material properties
Embed/enclose	Embed functional material
	Embed functional component
	Use enclosed, functional parts
Lightweight	Replace internal structure with lightweight cellular/lattice structure
	Hollow out artifact to reduce weight
Material distribution	Use material distribution to achieve desired behavior
	Absorb energy with small-interconnected parts
	Allow movement with small-interconnected parts
	Remove material to provide function
	Optimize structural topology or geometry
Reconfiguration	Create multifunctional artifact with reconfigurable structures

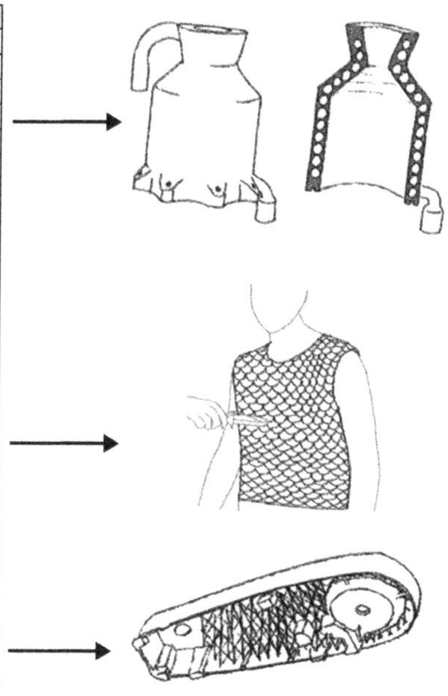

Figure 3.2 Categorization of design heuristics and some examples.

time, increased robustness, and multiple functions. For instance, the fuel burner can integrate internal channels for cooling fluid for higher combustion temperature. The full list of design heuristic cards can be downloadable here.[1] In a typical design heuristic card (see Figure 3.3a), it will contain the following information: 1) design heuristic (Tier II), 2) design heuristic card number, 3) description of design heuristic, and 4) design heuristic category (Tier I). In the ideation process, designers can browse the list of design heuristic cards for inspirations. User studies have shown that the developed heuristic cards are effective in increasing novice designers' design creativity [4]. Design heuristics in the form of physical object which is 3D printed to demonstrate the various AM freedoms. As shown in Figure 3.3b, multi-colored parts can help to show the novel attributes of AM. Compared to the pure usage of heuristic cards, the combination of cards and physical objects can outperform the former in terms of understanding of AM content, increasing novelty of generated solutions, and variety [15]. The physical heuristic models are downloadable on Thingiverse.[2] A redesign example of a bike is shown in Figure 3.4. By using the design heuristics, various design opportunities that are inspired by the heuristics can be quickly identified.

1 Heuristic cards. https://www.research-collection.ethz.ch/handle/20.500.11850/537327
2 Heuristic objects: https://www.thingiverse.com/edac_eth_zurich/collections/design-heuristics-for-additive-manufacturing

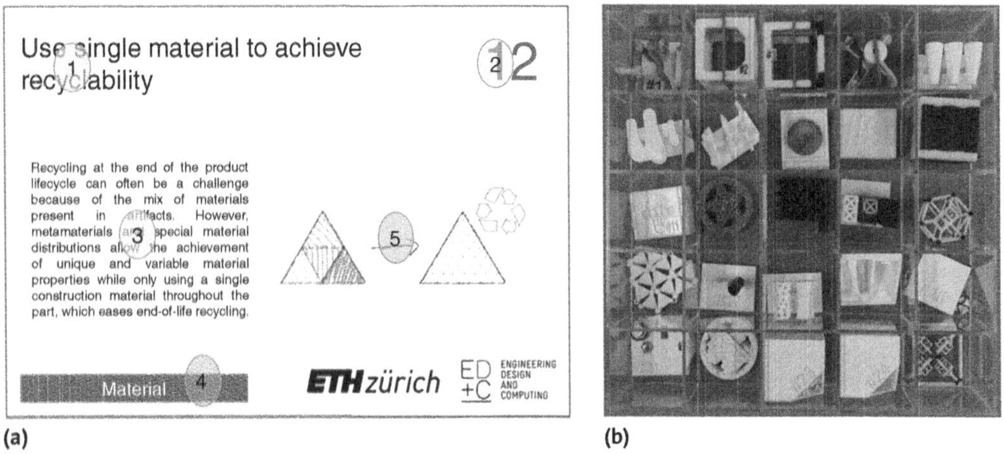

Figure 3.3 Design heuristics: (a) an example of design heuristic card and (b) design heuristic object. *Source*: [15] / with permission of American Society of Mechanical Engineers (ASME).

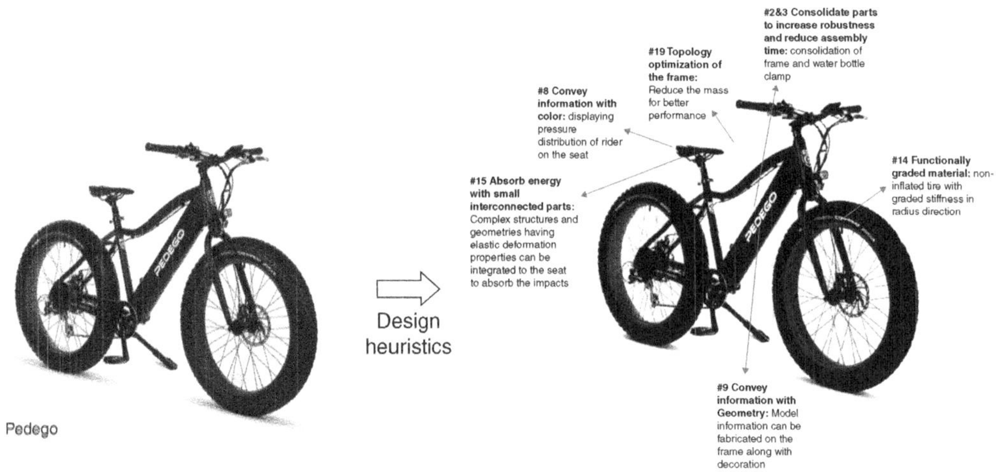

Figure 3.4 Application of the design heuristics for the redesign conceptualization of a bike.

3.4 Integrated DfAM Method

While various design methods have been reported for AM, these methods are highly scattered and focused on different design stages. The lack of an integrated DfAM framework with well-defined interfaces with existing design for X (DfX) approaches makes it difficult for industrial applications. Designers from industry who have received training of DfX may feel hesitated to adopt AM. To cope with this challenge, Kumke et al. [5] developed a framework to integrate both traditional design methods and newly developed DfAM methods using the general VDI 2221 model [16]. This framework can help designers to quickly build up their existing design knowledge and identify new design solutions with AM-specific knowledge. As shown in Figure 3.5, the framework has three main phases: planning and task clarification, conceptual design, and embodiment design. In the task clarification phase, product requirements are compiled. In this phase, existing general tools for supporting task clarification can be employed. Such tools include brainstorming, quality function deployment, function–behavior–structure, and TRIZ (theory of inventive problem solving), etc. In the conceptual design stage, AM-specific methods can be integrated into the framework. After breaking down the functions and generating the initial structure, AM-specific potentials can be exploited through topology optimization, design heuristics, and DfAM worksheet. Various conceptual design solutions will be evaluated based on manufacturing feasibility, economic constraints, and functionality. AM-specific constraints may also apply at this stage, for example, feasible volume that the 3D printer allows. Failure to meet these constraints will lead of disposal of the design solution or decomposition of the part into smaller subparts. In the embodiment and detailed design phase, optimization techniques can be applied such as topology optimization for lightweight structure, computation fluid dynamics for fluid-structure interactions, and acoustic optimization. Then design rules regarding AM process-specific constraints including support structure, surface roughness, material properties, etc. will help to secure the manufacturability of the optimized parts. Finally, digital validation of the optimized design solution should be implemented to satisfy functional, economic, and safety requirements. Through the above steps, a feasible solution with imparted AM potentials while satisfying the process constraints will be synthesized.

An example of assembly redesign using the framework is shown in Table 3.2. The housing cover is comprised of three main parts and eight screws. The first two steps are understanding the design requirements and mapping the functions to structures. The goal of redesign is to reduce the number of parts and create lightweight structures. In step 3, the possibility of having lattice structure to reduce the weight has been identified by the novice designer using tools such as AM feature database [7]. In the meantime, the possibility of having part consolidation to reduce the number of parts is also identified. The consolidated design solution can potentially eliminate leakage. In step 4, no further breakdown of the one integrated design is needed. In step 5, AM process selection is conducted to identify the best process considering the requirements of volume size, material properties, and quality. In this example, selective laser melting (SLM) process is selected. In step 6, the economic batch size of one makes customization of medical parts feasible. In step 7, no further optimization task takes place, such as topology optimization or CFD, because the lack of experience for the novice designer. Step 8 is to evaluate the manufacturing feasibility of the consolidated part with lattice infill. The constraints of SLM process are examined by using design rules. Finally, finite element simulation is carried out to check the pressure resistance of the redesign part. It should be noted that using this methodological framework alone is insufficient to support design for AM. More importantly, this framework offers a systematic way of applying various existing methods and AM-specific methods. The details of AM-specific methods such as part consolidation, topology optimization, lattice design, and design rules will be introduced in other chapters of this book.

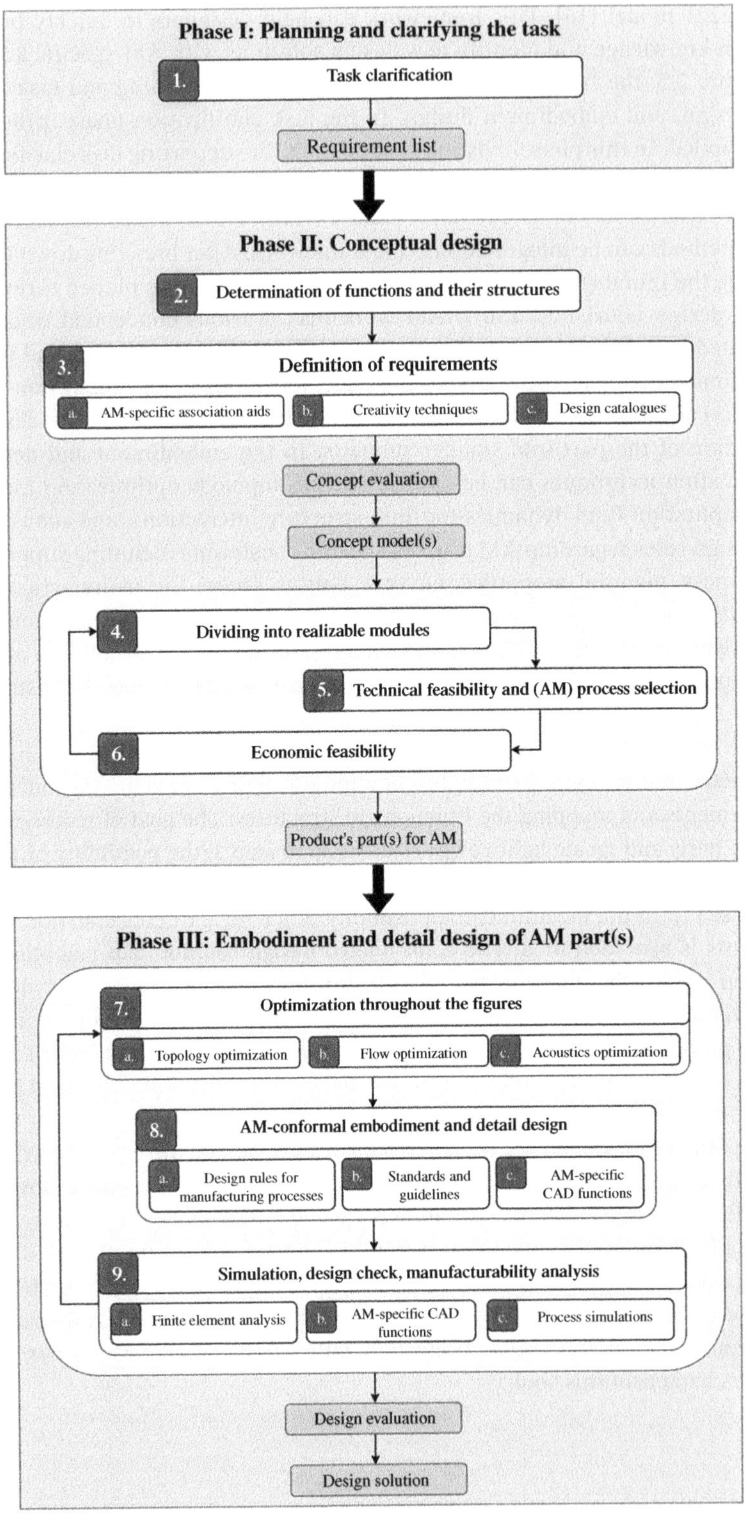

Figure 3.5 Integrated Design for AM framework. *Source*: Adapted from [5].

Table 3.2 A redesign example of a housing cover assembly.

Input	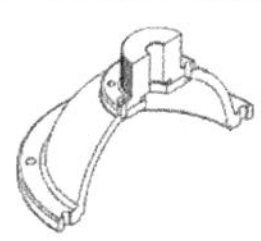 Product name: Housing cover Main parts: housing cover, threaded socket, 8 screws, O-ring
Step 1	**Task clarification and definition of requirements**: • Requirements of previous part serve as basis • User adds new requirements based on design goals and compiles requirements list
Step 2	**Determination of functions and their structures** • This step is skipped (function structures are given due to redesign)
Step 3	**Development of basic solution ideas** • User uses bionics catalogue and finds bones/honeycombs as ideas for lightweight design • User chooses to incorporate lattice structures enabled by AM Am feature database redirects user to module F to investigate AM-enabled parts consolidation
Step 4	**Functional extension and part consolidation** • Part consolidation principles provide the idea that a lower number of joints can reduce weight • User analyzes neighbor parts of assembly and finds that the threaded socket could be included into the housing cover • User identifies positive side effect: eliminated joint can also prevent potential leakage and thus increase reliability **Dividing into realizable modules** • User decides to turn former assembly into single part
Step 5	**Technical feasibility and AM process selection** • User utilizes decision support system and finds that only selective laser melting provides sufficient material characteristics, accuracy, and build chamber size
Step 6	**Economic feasibility** • User finds that economic feasibility is given due to small lot size and reduced assembly cost
Step 7	**Optimization of specific product properties** • Module is skipped (novice user has no experience in optimization tools)
Step 8	**AM-conformal embodiment and detail design** • User uses AM design rules catalogue to generate AM-conformal design • User finds suitable lattice structures for support-free manufacturing • AM-specific CAD functions provide assistance • User finds that surface quality and accuracy are not sufficient for manufacturing threads and therefore also has to adhere to DFM guidelines for machining
Step 9	**Simulation, design check, manufacturability analysis** • User carries out finite element analysis to check pressure resistance • User has no access to tools for design check and thus asks AM service provider for manufacturability analysis
Output	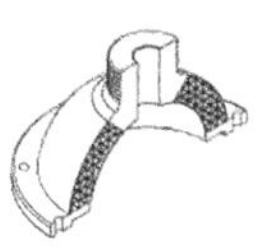 New design solution: Housing cover with integrated threaded socket • Lower weight due to lattice structures • Improved reliability due to part consolidation (reduced risk of leakage)

Source: Adapted from [5].

3.5 Automated AM Candidacy Prediction Tools

Identification of redesign potentials of parts or assemblies that would benefit from adopting an AM solution is one of the main tasks in design for AM. However, identifying the right adoption strategy remains challenging, as no universal formula guarantees successful implementation. Success largely depends on how well a company's specific needs align with the unique capabilities of AM. While certain restrictive guidelines—such as limits on build volume and batch size [17]—have been proposed to support decision-making, the broader trade-offs between these constraints and the potential geometric, functional, economic, and societal benefits are less frequently explored. To solve the problem of identifying part candidacy, a few methods, including heuristic ones [14, 18, 19] and computational ones [20–22], have been reported. Current methods tend to rely heavily on expert input or concentrate on a narrow set of AM benefits, such as weight reduction or component integration. As the number of parts under consideration expands—particularly in large or complex assemblies—efficiency and computational cost become major challenges. This highlights the need for an automated approach that can swiftly pinpoint suitable AM candidates and indicate the potential advantages AM might offer. Such a tool could provide an initial screening stage for nonspecialist users (e.g., product managers), allowing them to focus on a smaller subset of parts before seeking expert evaluation. This section will introduce an automated tool to support the identification of AM part candidates using supervised machine learning (ML) [23].

3.5.1 AM Candidacy Criteria

AM facilitates a wide range of benefits ranging from economic batch size to design potentials. In conceptual design stage, enlarging the design solution space with the aid of a computational tool is more favored. As such, the AM candidacy criteria will mainly focus on those that are computational and easily accessible. The following two "filtering principles" are used to down-select the candidacy list that is surveyed from literatures and AM artifacts [23].

- **Principle of Efficiency**: How can the data required to evaluate and justify AM candidacy be obtained and processed with minimal effort?
- **Principle of Objectivity**: Can the criteria for determining AM suitability be assessed in a quantifiable and unbiased manner?

The sampled AM candidacy criteria is summarized in Table 3.3. The list of criteria is not exhaustive. Additional factors that align with the principles of efficiency and objectivity may be incorporated in future studies. Moreover, these criteria are parameterized for easy determination of suitability for AM design. For instance, in geometric analysis, traditional feature-based complexity measures often rely heavily on manual interpretation and may fail when dealing with compound features [24]. Consequently, feature-independent computational metrics—such as volume-to-surface ratio, surface-to-bounding-box (BB) volume ratio, and volume-to-BB volume ratio—are generally preferred [25]. Model-level analysis can further support evaluating part consolidation potential by considering attributes like the number of components, fasteners, and assembly interfaces, which collectively indicate assembly complexity. A higher level of complexity typically suggests greater potential for consolidation. The parameters used for such evaluations can be sourced from users, computer-aided design (CAD) models, and enterprise resource planning (ERP) systems, enabling automated extraction of relevant information.

Table 3.3 Selected candidacy criteria and its key parameters.

Candidacy criteria	Decisive parameters	#	Units	Source
	Ratio of part volume to part surface area	DP1		CAD
Geometric analysis	Ratio of part surface area to bounding box volume	DP2		CAD
	Ratio of part volume to bounding box volume	DP3		CAD
	Number of components	DP4		CAD
Model analysis	Number of fasteners	DP5		User
	Number of assembly interfaces	DP6		User
	Manufacturing cost	DP7	$/part	ERP
	Batch size	DP8	# of parts	ERP
Economic analysis	Lead time	DP9	days	ERP
	Inventory costs	DP10	$	ERP
	Imports/exports cost	DP11	$	ERP
	Does the part contain internal channels/structures?	DP12	Y/N	User
Design potential analysis	Does the part have any surface markings?	DP13	Y/N	User
	Are there similar parts with similar modifications?	DP14	Y/N	User
	Does the part require human body compliance?	DP15	Y/N	User

3.5.2 Machine Learning-assisted AM Candidacy Decision Framework

The ML-assisted AM candidacy decision framework is shown in Figure 3.6. The framework is comprised of three main branches. The first branch is to derive the AM candidacy criteria based on a pool of surveyed AM potentials and the filtering principles that are discussed in Section 3.5.1. The second branch of the framework aims to extract data from historical records for training the ML model. The training dataset is compiled from multiple sources, including non-AM cases (e.g., components produced via CNC milling), successful AM implementations, and failed AM attempts. Each case is assessed by AM experts based on the defined candidacy criteria and annotated with the specific AM potentials it exhibits. The final component of the framework involves developing decision models that leverage these labeled examples to predict the AM suitability of new applications. The model's output provides a percentage-based estimation of potential AM benefits. Ultimately, each newly evaluated case is added to the labeled database, enabling continuous refinement and improved accuracy of the predictive model over time.

Extraction and labeling of training data are critical for the success of the decision model. In this study, over 200 parts are analyzed regarding the 15 decisive parameters under the categories of geometry, model, economy, and design potentials. The economic evaluation is conducted through the Xometry quoting plugin [26], an online manufacturing service platform that supports various production processes, including 3D printing and CNC machining. This tool provides rapid estimates of lead time and manufacturing cost with an approximate ±5% accuracy margin. To ensure data integrity and labeling consistency in AM suitability grading, two AM experts—each with over three years of experience in product design and AM-focused industrial projects—jointly assess each instance. A consensus-based approach is adopted to determine the presence of specific AM potentials. Figure 3.7 illustrates a representative labeled

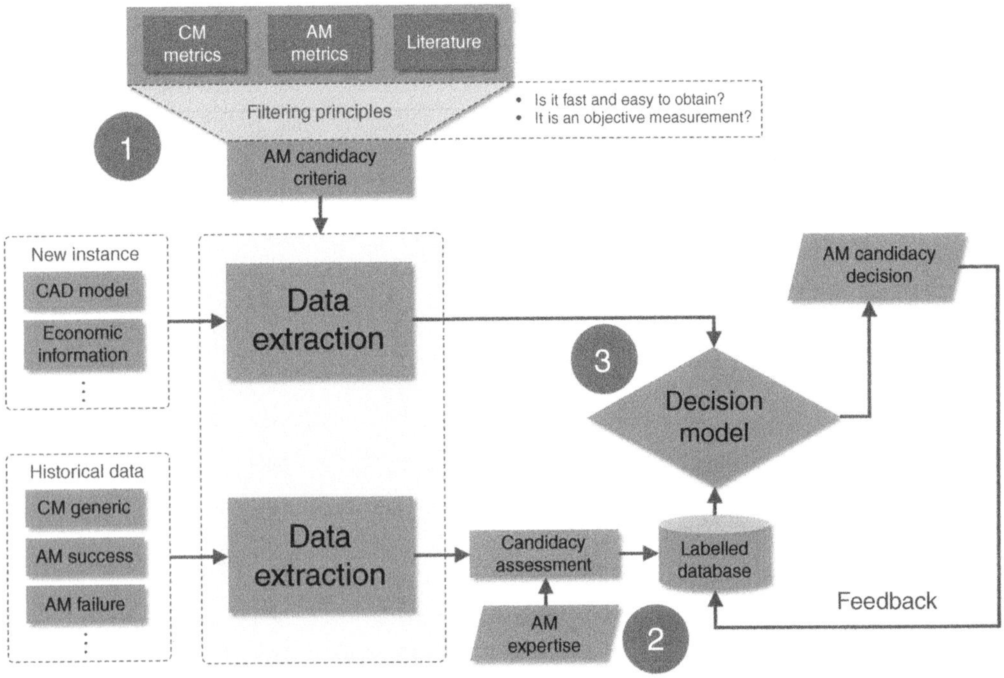

Figure 3.6 The proposed framework of the decision support system for AM part candidacy identification.

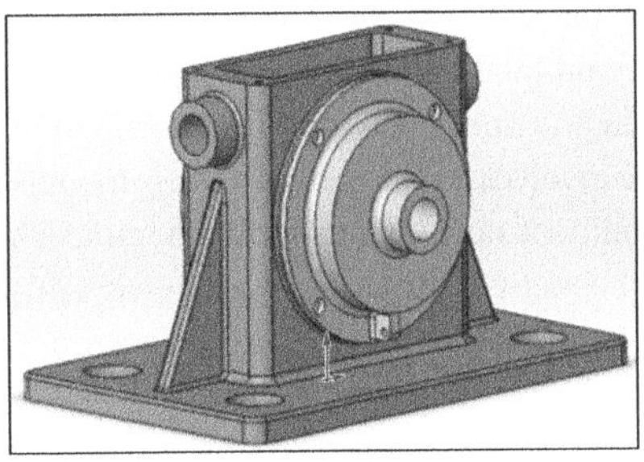

(a)

COLUMN_NAME	IS_NULLABLE	DATA_TYPE	Example Assembly Values
id	NO	varchar	ee51hdb-183bjda-ffahs2
volume	YES	float	614489.2
surface_area	YES	float	154840.06
bounding_box_x	YES	float	203.2
bounding_box_y	YES	float	127
bounding_box_z	YES	float	120.65
components	YES	int	3
fasteners	YES	int	8
assembly_interfaces	YES	int	2
cost	YES	float	183.68
batch_size	YES	float	30
lead_time	YES	float	15
inventory_cost	YES	float	0
import_export_cost	YES	float	0
surface_markings	YES	bool	FALSE
internal_channels	YES	bool	FALSE
similar_parts	YES	bool	FALSE
size_variation	YES	bool	FALSE
human_body	YES	bool	FALSE
result_eco_feas	YES	float	1.0
result_lightweight	YES	float	1.0
result_pc	YES	float	1.0
result_custom	YES	float	0.0
result_internal	YES	float	0.0

(b)

Figure 3.7 An example of labeled instance with input parameters and identified AM potentials. (a) Gearbox assembly and (b) decisive parameters and identified potentials.

instance, showing both the input parameters and the identified AM potentials, each encoded using a binary classification scheme. While minor subjectivity may persist in expert labeling, the integration of real industrial data from enterprise ERP systems is expected to strengthen the dataset's reliability and contribute to more robust and data-driven decision-making in future model iterations.

The identified decisive parameters are used as training features for developing the ML model. Five commonly used regression algorithms were initially evaluated for each AM potential, namely linear regression, Bayesian linear regression, neural network regression, boosted decision tree regression (BDTR), and decision forest regression. The root-mean-square error (RMSE) was employed as the performance metric. A training set of 200 samples was used, with 10% reserved for cross-validation. Among the tested models, the BDTR algorithm achieved the lowest RMSE across all AM potentials, outperforming the other approaches. This superior performance can be attributed to BDTR's ability to capture complex nonlinear relationships and its robustness in small-sample learning environments [27–29]. Consequently, BDTR was selected as the primary ML algorithm for predicting AM candidacy in the proposed framework.

3.5.3 A Cloud-based Tool Based on the Decision Framework

The proposed AM candidacy prediction framework is implemented as a cloud-based application, designed for broad accessibility and real-time updating of both the training dataset and the trained models. The developed tool is available through the project website [30], and its open-source source code can be accessed via GitHub [31].

A range of web technologies supports the framework to ensure high functionality, maintainability, and scalability. The communication structure among these technologies, shown in Figure 3.8, forms the core architecture of the tool. Because several third-party services are integrated, PHP (Hypertext Preprocessor) is used as an intermediary to manage communication between the various APIs (Application Programming Interfaces). This approach conceals API keys and avoids the access restrictions typically associated with HTTP (HyperText Transfer Protocol). Within the system, Autodesk Forge is primarily responsible for CAD file conversion (e.g., from SolidWorks to STL), metadata extraction through the Model Derivative API, and remote storage of CAD models. Microsoft Azure handles training database management and ML operations through API integration. Meanwhile, AJAX (Asynchronous JavaScript and XML) enables dynamic, real-time communication between the client and server without requiring

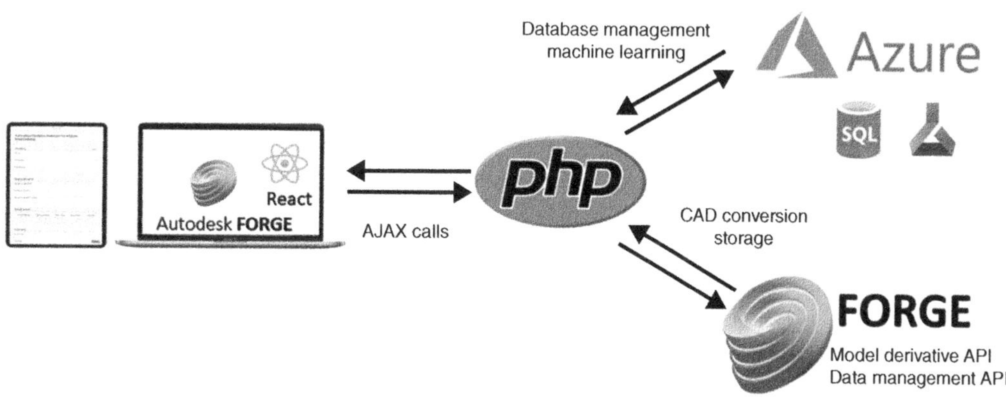

Figure 3.8 The architecture of implementation of the AM candidacy prediction framework.

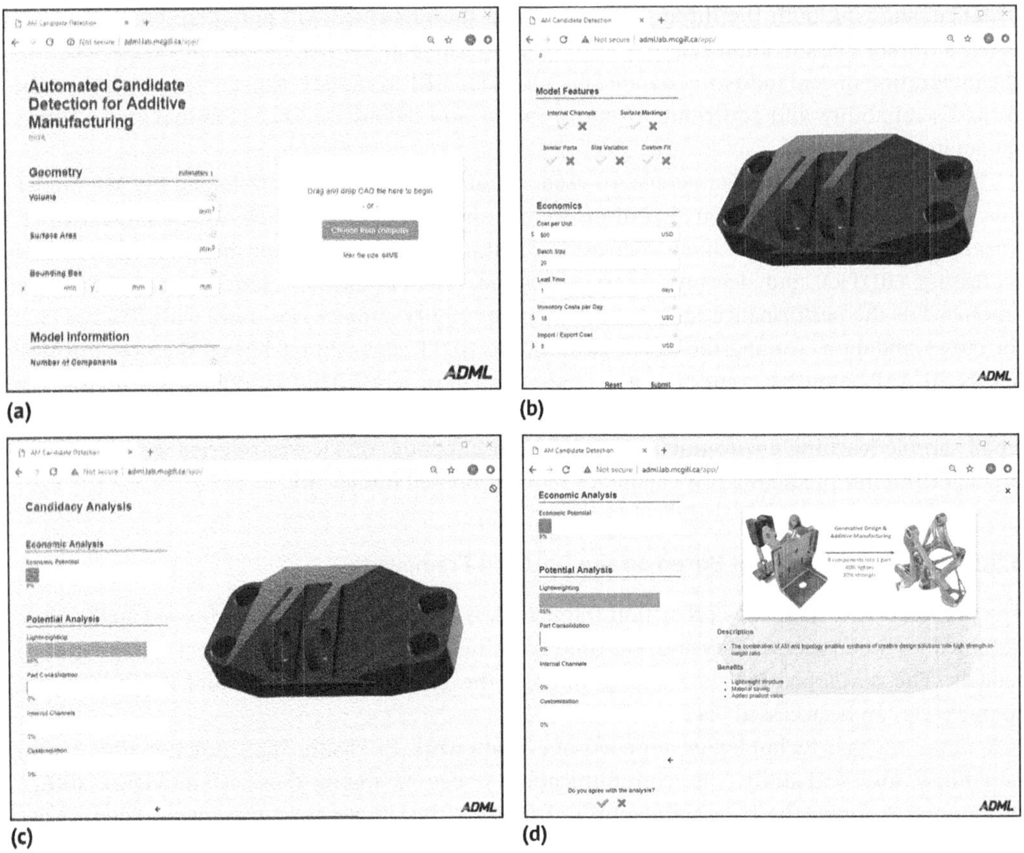

Figure 3.9 A collection of snapshots of the developed tool. (a) Load CAD model, (b) data acquisition, (c) candidacy prediction, and (d) lightweighting potential tutorials.

a page refresh. The user interface is developed using the React JavaScript library [32], which allows for an interactive, multi-state single-page application that delivers a responsive and seamless user experience.

The user interface of the prediction tool is shown in Figure 3.9. The prediction process starts with loading a CAD model and automatically extracts geometric information, such as volume, surface area, and number of components. Users are then prompted to complete any missing information related to the model, such as the number of interfaces, specific geometric features (e.g., the presence of internal channels), and economic parameters (e.g., estimated lead time). It is worth noting that, in future iterations, many of these manual inputs could be automated through integration with an ERP system. Once all inputs are provided, the tool generates candidacy scores for each AM potential, as illustrated in Figure 3.9c. To support user learning, the system includes interactive tutorials accessible by hovering over specific items (e.g., the "lightweight" bar in Figure 3.9d). These tutorials aim to help novice AM users better understand the potential benefits of adopting AM, aligning with observations from prior research [3]. To enhance the framework's predictive accuracy, the tool incorporates a feedback mechanism allowing users to determine whether a newly evaluated instance should be added to the training dataset for future model improvement. If deemed irrelevant, the instance can be discarded to maintain data quality.

3.5.4 Case Study

Two examples of a hip implant [33] and a throttle pedal [21] are presented to demonstrate the efficacy of the candidacy prediction framework as well as the cloud-based decision tool. These case studies were selected because their AM potentials have been previously examined in the literature, providing a means to cross-validate the tool's predictive results. The details of the examples are summarized in Table 3.4. The hip implant model [34], includes cost-related data estimated from online sources. As shown in Figure 3.10a, the predicted outcomes indicate high potentials for economics (100%), customization (99%), and light-weighting (98%), while the potentials for part consolidation (24%) and internal channels (0%) are low. These results are consistent with findings reported by Fraunhofer IWU [35], where the MUGETO implant employed lattice structures to achieve weight reduction. The predicted economic advantage primarily stems from low production volume, while the one-of-a-kind customization and anatomical conformity highlight its high customization potential. However, due to the limited number of parts and interfaces, the part consolidation potential requires further evaluation using a specialized verification tool, such as the one developed by Yang et al. [20]. Similarly, the throttle pedal model [36] uses cost data estimated from publicly available information. The predicted results, shown in Figure 3.10b, suggest high potentials for part consolidation (100%) and light-weighting (76%), but lower scores for economics (28%), customization (0%), and internal channels (0%). These findings align closely with those reported study [20], further validating the reliability of the proposed prediction framework.

Table 3.4 Details of the tested examples.

Name	Figures	Volume (mm^3) Area (mm^2) BD box (mm^3)	# Components # Fasteners # Interfaces	Cost Quantity Lead time Inventory Import/export	Markings Channels Similar parts Size variance Human custom fit
Hip implant		55910.32 12367.67 166.38*106.46*39.97	2 0 1	2000 1 1 0 0	0 0 0 0 1
Throttle pedal		210367.17 112685.27 156.22*218.77*56.48	41 17 15	350 10000 1 2 0	0 0 0 0 0

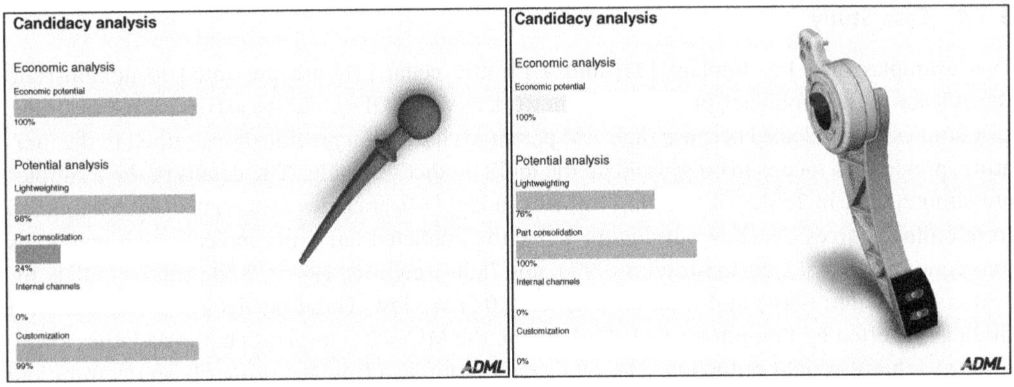

Micro Technologies, S.A.

(a) (b)

Figure 3.10 AM candidacy prediction of the hip implant and the throttle pedal. (a) Potential prediction of hip implant and (b) potential prediction of throttle pedal.

3.6 Multifunctional Bio-inspired Design Method

Multifunctional bio-inspired design (MBID) [37–39] is a systematic ideation framework that supports the generation of bio-inspired concepts tailored for AM. The method begins with functional decomposition of design requirements, identifying both primary and secondary functions desired in the final product. Designers then retrieve biological analogies using databases and targeted research, mapping biological strategies onto engineering needs through function–structure–behavior (FSB) translation. The framework integrates this mapping into morphological charts or concept matrices that enable combinatorial exploration of multifunctional solutions.

MBID's structured approach is particularly well suited to opportunistic DfAM, where the aim is not only to make parts manufacturable but to rethink them entirely using nature's efficient, integrated strategies. By encouraging designers to abstract engineering problems in functional terms and match them with biological precedents, MBID opens up design spaces that would be difficult to explore using traditional heuristics alone. It is especially effective in early-stage design, where ambiguity is high and innovation is needed. While MBID is not a ML tool, it exemplifies a data-informed ideation method that supports AM's capability to produce structurally complex, multifunctional parts.

3.7 Summary

Conceptualization is the starting point of employing AM freedoms into the product design journey. Being exposed to the novel design potentials and constraints of AM, designers can potentially identify new design solutions earlier while being aware of the feasible constraints. In this chapter, five different approaches have been introduced as viable routes for design conceptualization with AM. There is no superior advantage of one approach over another, and these methods can be combined to some extent. For example, the integrated DfAM method can serve as a guideline for organizing design activities. The automated AM candidacy prediction tool can be used as a fast-screening tool for down-sampling a large set of parts regarding what potentials a given part can pursue.

The design heuristics method can serve as a more detailed description of AM potentials with respect to what it is and how to implement it in a design. For example, from the automated AM candidacy prediction tool, a legacy part is dragonized with 80% potential for lattice structure for lightweight. The design heuristics card on lattice structure can show examples of lattice and inspire ideas of lattice creation. Finally, DfAM worksheet can help to have rough estimation whether the part is feasible for AM or not. It should be advised that conceptualization process is highly intuitive and knowledge-intensive such that domain knowledge of a given design still plays a dominant role in generating design solutions. The methods introduced in the chapter should be counted as sources of inspiration.

References

1 Thompson, M.K., et al., *Design for additive manufacturing: trends, opportunities, considerations, and constraints*. CIRP Annals-Manufacturing Technology, 2016. **65**(2): pp. 737–760.

2 Yang, S. and Zhao, Y., *Additive manufacturing-enabled design theory and methodology: a critical review*. The International Journal of Advanced Manufacturing Technology, 2015. **80(1)**: pp. 327–342.

3 Yang, S., Page, T., and Zhao, Y.F., *Understanding the role of additive manufacturing knowledge in stimulating design innovation for novice designers*. Journal of Mechanical Design, 2019. **141**(2): p. 021703.

4 Blösch-Paidosh, A. and Shea, K., Design heuristics for additive manufacturing. in *Proc. DS 87-5 Proceedings of the 21st International Conference on Engineering Design (ICED 17) Vol 5: Design for X, Design to X, Vancouver, Canada, 21–25.08*, 2017.

5 Kumke, M., Watschke, H., and Vietor, T., *A new methodological framework for design for additive manufacturing*. Virtual and Physical Prototyping, 2016. **11**(1): pp. 3–19.

6 Rosen, D.W., *A review of synthesis methods for additive manufacturing*. Virtual and Physical Prototyping, 2016. **11(4)**: pp. 305–317.

7 Bin Maidin, S., Campbell, I., and Pei, E., *Development of a design feature database to support design for additive manufacturing*. Assembly Automation, 2012. **32**(3): pp. 235–244.

8 Dinar, M. and Rosen, D.W., *A design for additive manufacturing ontology*. Journal of Computing and Information Science in Engineering, 2017. **17**(2): pp. 021013-021013-021019.

9 Richter, T., et al., Exploitation of potentials of additive manufacturing in ideation workshops. in *The Fifth International Conference on Design Creativity (ICDC2018)*. Bath, United Kingdom. 2018.

10 Kumke, M., et al., *Methods and tools for identifying and leveraging additive manufacturing design potentials*. International Journal on Interactive Design and Manufacturing (IJIDeM), 2017. **12(2)**: pp. 481–493.

11 Floriane, L., et al., *Enriching design with X through tailored additive manufacturing knowledge: a methodological proposal*. International Journal on Interactive Design and Manufacturing (IJIDeM), 2016. **11(2)**: pp. 279–288.

12 Sinha, S., et al., Does designing for additive manufacturing help us be more creative? an exploration in engineering design education. in *Proceedings of the ASME 2017 International Design Engineering Technical Conferences and Computers and Information in Engineering Conference*, American Society of Mechanical Engineers, 2017. pp. V003T004A014–V003T004A014.

13 Booth, J.W., et al., *The design for additive manufacturing worksheet*. Journal of Mechanical Design, 2017. **139**(10): p. 100904.

14 Fu, K.K., Yang, M.C., and Wood, K.L., *Design principles: literature review, analysis, and future directions*. Journal of Mechanical Design, 2016. **138**(10): p. 101103.

15 Blösch-Paidosh, A. and Shea, K., *Enhancing creative redesign through multimodal design heuristics for additive manufacturing*. Journal of Mechanical Design, 2021. **143**(10): p. 102003.

16 und Mechatronik, V.F.P., *Systematic approach to the development and design of technical systems and products*. 1993, Beuth Verlag GmbH.

17 Doubrovski, Z., Verlinden, J.C., and Geraedts, J.M., Optimal design for additive manufacturing: opportunities and challenges. in *Proceedings of the ASME 2011 International Design Engineering Technical Conferences and Computers and Information in Engineering Conference*, American Society of Mechanical Engineers, 2011. pp. 635–646.

18 Klahn, C., Leutenecker, B., and Meboldt, M., *Design for additive manufacturing–supporting the substitution of components in series products*. Procedia Cirp, 2014. **21**: pp. 138–143.

19 Reiher, T., et al., *Holistic approach for industrializing AM technology: from part selection to test and verification*. Progress in Additive Manufacturing, 2017. **2(1)**: pp. 43–55.

20 Yang, S., Santoro, F., and Zhao, Y.F., *Towards a numerical approach of finding candidates for additive manufacturing-enabled part consolidation*. Journal of Mechanical Design, 2018. **140**(4): pp. 041701-041701-041713.

21 Yang, S., et al., *A numerical-based part consolidation candidate detection approach with modularization considerations*. Research in Engineering Design, 2019. **30**(1): pp. 63–83.

22 Yao, X., Moon, S.K. and Bi, G., *A hybrid machine learning approach for additive manufacturing design feature recommendation*. Rapid Prototyping Journal, 2017. **23**(6): pp. 983–997.

23 Yang, S., et al., *Towards an automated decision support system for the identification of additive manufacturing part candidates*. Journal of Intelligent Manufacturing, 2020. **31**(8): pp. 1917–1933.

24 Babic, B., Nesic, N., and Miljkovic, Z., *A review of automated feature recognition with rule-based pattern recognition*. Computers in industry, 2008. **59**(4): pp. 321–337.

25 Valentan, B., et al., *Basic solutions on shape complexity evaluation of STL data*. Journal of Achievements in Materials and Manufacturing Engineering, 2008. **26**(1): pp. 73–80.

26 Xometry. *Instant quoting add-in for SOLIDWORKS and autodesk inventor*. 2017, https://www.xometry.com/cad-add-in-downloads.

27 Roe, B.P., et al., *Boosted decision trees as an alternative to artificial neural networks for particle identification*. Nuclear Instruments and Methods in Physics Research Section A: Accelerators, Spectrometers, Detectors and Associated Equipment, 2005. **543**(2–3): pp. 577–584.

28 Coadou, Y., Boosted decision trees and applications. in *Proceedings of the EPJ Web of conferences*, EDP Sciences, 2013. 55: p. 02004.

29 Xia, Y., et al., *A boosted decision tree approach using Bayesian hyper-parameter optimization for credit scoring*. Expert Systems with Applications, 2017. **78**: pp. 225–241.

30 (ADML), A. D. a. M. L. *Automated candidate detection for additive manufacturing (BETA)*. 2019, http://adml.lab.mcgill.ca/app/.

31 (ADML), A. D. a. M. L. *ADML website app*. 2019, https://github.com/adml-mcgill/website/tree/master/app.

32 OpenSouce, F. *React: a JavaScript library for building user interfaces*. 2019, https://reactjs.org/.

33 Ryan, G., Pandit, A., and Apatsidis, D.P.J.B., *Fabrication methods of porous metals for use in orthopaedic applications*. Biomaterials, 2006. **27**(13): pp. 2651–2670.

34 Fuentes, E., *Hip replacement prosthesis*. 2012, https://grabcad.com/library/hip-replacementprosthesis.

35 Schnabel, T., et al., *Design for additive manufacturing-guidelines and case studies for metal applications.* Procedia Manufacturing, 2017. **11**: p. 141283.

36 Miessner, H. *Throttle pedal design challenge.* 2015, https://grabcad.com/library/pedal-one-microtechnologies-1.

37 Velivela, P.T. and Zhao, Y.F., *Supporting multifunctional bio-inspired design concept generation through case-based expandable domain integrated design (xDID) model.* Designs, 2023. **7**(4): p. 86.

38 Velivela, P.T. and Zhao, Y.F., *BIKAS: bio-inspired knowledge acquisition and simulacrum—a knowledge database to support multifunctional design concept generation.* Data Intelligence, 2024. **6**(2): pp. 531–558.

39 Velivela, P.T., Ridard, A., and Zhao, Y.F., *Parameters for selecting biological features in multifunctional bio-inspired design: a convergent evolution approach.* Bioinspiration & Biomimetics, 2024. **19**(3): p. 036020.

4

Part Consolidation and Assembly Design

This chapter introduces the concepts of part consolidation (PC) and assembly-free design that reduces part count. The reduced part count typically makes the geometry more complicated, thereby increasing the difficulty of manufacturing. Therefore, a trade-off has to be made regarding 1) which parts to be consolidated, and 2) how to consolidate the parts. Various rules and tools for supporting PC decisions are entailed in this chapter.

4.1 What Is Part Consolidation?

PC is one of the core concepts in design for assembly [1] to simplify product architecture, achieve integrated functions, reduce part count, eliminate the need of assembly, and avoid fatigue issues due to poor assembly quality. PC also provides downstream benefits including inventory reduction, less tooling, cost reduction, and lead time reduction. Given its numerous benefits, PC has been widely practiced in industry over the past three decades. However, there is no clear definition of PC in literature. According to the Oxford English Dictionary, "consolidation" is the action or process of combining a number of things into a single more effective or coherent whole [2]. Hereby, PC can be defined as "the act of combining several parts in an assembly into a single part that can be easily manufactured." For the ease of reference, the new design after PC is termed as consolidated design in contrast with the assembly design. Given the long history of PC research, we divide the research on PC by two streams: PC in conventional manufacturing (CM) context and PC in AM context.

4.1.1 PC in CM Context

PC in CM context are normally studied under the philosophy of design for manufacturing and assembly (DfMA) [1, 3], which aims at examining the effect of design on a production system (or vice versa). Compared to assembly design counterpart, PC design technique has reported to help IBM® to increase productivity by 700% after reducing part count by two-thirds. Similarly, Ford® has benefited from reduced part count in its door trim by 79%, assembly cost by 94%, and material costs by 27%.

The majority of works on PC serve as a proof of concept and rely heavily on the experience of design experts. The top challenges of applying AM-PC potentials can be summarized as follows:

- **Part candidate selection**. Given a complex assembly, how to efficiently filter parts that are feasible for PC?
- **Fusion of parts**. Given the filtered part candidates, how to convert the assembly design to a consolidated design with full consideration of AM design freedoms and constraints?

To shed a light on these challenges, this chapter is organized as follows. Research literature on selection of part candidates for consolidation and conversion of assembly design to consolidated design are firstly summarized. Then, a holistic approach for supporting PC design is introduced with integrated modules of part filtering and fusion of parts in Section 4.3. Details of the key techniques on the two modules are introduced in Sections 4.4 and 4.5, respectively, with a gas pedal example. In the end, some emerging trends in PC research are discussed.

4.2 Literature Review

4.2.1 Selection of Part Candidates for Part Consolidation

This section reports the rules and methods in literature that help to filter PC candidates in both AM and CM contexts. A comprehensive set of design rules related to PC decisions was summarized in literature [13] and cited as follows:

1. **Decoupling of functions (r1)**: Functions should not have overlapped physical design parameters as suggested by axiomatic design [14] and TRIZ [15]. The coupled design may have issues of increased development time and conjugate failure. Therefore, components to achieve multiple functions need special attention.
2. **No relative motion (r2)** [1]: If there exists relative motion between components, PC is not suggested.
3. **No material variance (r3)** [1]: If components are made of different materials, separation is favored.
4. **No blockage of assembly access of other components (r4)** [1]: If consolidation jeopardizes the assembly access of other components, consolidation is forbidden.
5. **No extra tooling (r5)** [16]: Consolidated part should not need extra tooling.
6. **No vast difference of quality requirements (r6)** [17]: Similar GD&T requirements simplify process routes.
7. **No significant increase of manufacturing cost (r7)** [17]: Consolidated part tends to be more geometrically complex, which may increase manufacturing cost.
8. **Trade-offs should be made between PC and standardization and modular design (r8)**: Consolidated parts are normally irregular, which is against modularization and standardization.
9. **Increase the possibility of part re-usage and material recyclability (r9)**: Similarity in terms of part lifespan and material recyclability helps to boost a more sustainable design. Design for disassembly [18] also benefits part re-usage and material recyclability.

Although these rules are widely practiced, some criticism is also received in case of some exceptions. For example, some parts with relative motion are still manufacturable if elastic deformation is permitted [17]. These rules show two main deficits: 1) rules r1–r9 work at the conceptual level without considering any specific manufacturing constraints; and 2) these rules overemphasize the feasibility and cost sides without touching the performance aspect, which leads to more conservative designs.

The advances in manufacturing capabilities of complex shapes and materials have sparked the rethinking of applying these design rules (r1–r9) to additive manufacturing context. According to the feasibility analysis by Yang et al. [13], all these rules are challenged by the expanded design freedoms of AM. Take r2 as an example, articulated mechanism (i.e., parts having relative motion) can be printed as a whole without any assembly operations. Similarly, r4 is challenged by AM, which can consolidate all parts including the ones that need assembly operations in CM context. However, as the authors stated that a new set of design rules that can take full advantage of AM design freedoms while securing quality of consolidated parts is in strong need.

Another important stream of part filtering for PC is the part candidate selection method. As shown in Figure 4.4, PC technique has been applied in optimizing the part count and architecture for products such as fuel nozzle [10], hydraulic manifold [19], air duct [11, 20], sawdust blower [21], motorcycle triple clamp [22], and engine [23]. The original number of parts ranged from 4 to 20 except one example which GE team successfully consolidated nearly 900 parts as 12 pieces for the helicopter engine prototype [23]. In most cases, the authors either did not mention the candidacy rules [10, 11, 20, 23] or simply used the obsolete heuristic rules of "no relative motion," "no material variance," and "no blockage of assembly access" [19, 22]. These rules, though they are helpful in down-sampling potential candidates, are not enough to support candidacy decision. For example, these heuristic rules did not take consideration of AM constraints such as available materials and allowable building volume. In addition, manufacturing constraints and design iterations are not detailed in these literatures. In conclusion, part filtering rules and part selection methods need to be revised to adapt to the expanded PC opportunities as well as be applicable to a more general and complex system instead of simple assemblies.

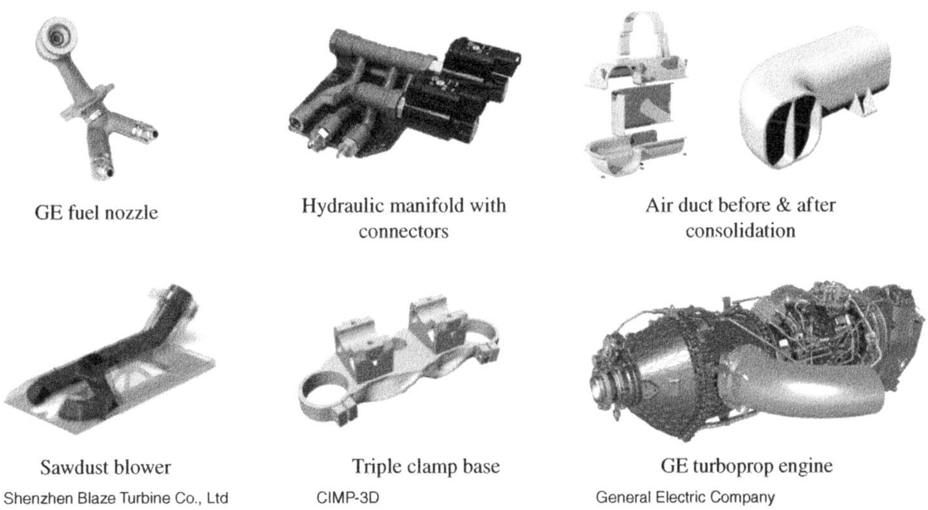

Figure 4.4 Some part consolidation examples reported in literature.

4.2.2 Transition From Assembly Design to Consolidated Design

On the successful identification of part candidates for PC, the next challenge is how to make the transition from the original assembly design to a consolidated design. Schmelzle et al. [19] proposed a redesign scheme for PC with three main steps including identification of candidate parts, specification of optimal external and internal geometry via simulation tools (e.g., finite element analysis [FEA] and computational fluid dynamics [CFD]), and iterative optimization of building orientation (BO), building support, and need of post-processing (e.g., support removal) to minimize building time, material usage, and distortion while achieving desired functionality. This approach provides a general design guideline of basic steps and important factors (e.g., BO) to be taken in the PC redesign process, but it fails to provide enough details of how to specify and consolidate external and internal geometric features. Ponche et al. [24] developed a functional surface (FS) and functional volume (FV)-oriented design method with full considerations of AM process characteristics (e.g., maximum and minimum dimensions, accuracy, and necessity of finish process). The detailed flow is shown in Figure 4.5. FS is derived from functional specifications and defined by designers. FV is the thickened FS in the normal direction and the thickness is determined by the minimum mechanical requirements (e.g., stress), layer thickness, and finish requirements. Once FVs are determined, linking volumes (LV) that connect discrete FVs are heuristically defined by the steps of 1) determining the proper BOs, 2) grouping FVs that can be built from the same substrate, 3) defining the shape of substrates and LVs. The proposal of FS, FV, and LV offers a novel perspective to define a design space, but this method falls short in the following aspects. First, how to define an FS is not provided. For example, an existing design is comprised of different types of features but not all of them are functional features (e.g., recessional undercut). Second, the shape and topology of truss-like LVs are highly dependent on designers' experience, which may not lead to the optimal LVs. An alternative solution of generating LVs could be the synthesis of a continuous encasing volume of all FVs and conducting topology optimization thereafter. Similar concepts of FS and FV are also reported in the literature [21, 25] to enable early determination of BO in the conceptual design stage. ASML® photolithography system applied the functional surface-based method to redesign a manifold assembly as shown in Figure 4.6a. Comparing with conventional design method of starting with a maximum volume and removing materials to form function features (e.g., channels) and reduce mass (e.g., lightweight holes), the FS-based method takes addition algorithm by defining FSs and maximum volume constraints and then adding functional features and reinforcement. Corresponding results are presented in Figure 4.6b and c. Orme et al. [26] adopted an enlarged solid model instead of using the directly consolidated entity as the initial design space (IDS) for topology optimization in a space engine case study, which leads to drastically different product structure. It highlights the importance of the definition of "right" IDS.

The overview also indicates the common deficiency of the existing methods showing no concern of design for function in the synthesis of FSs, although the original FSs are designed for CM. The other deficiency is the lack of considering other AM potentials besides PC such as light weighting and performance improvement after the consolidation process.

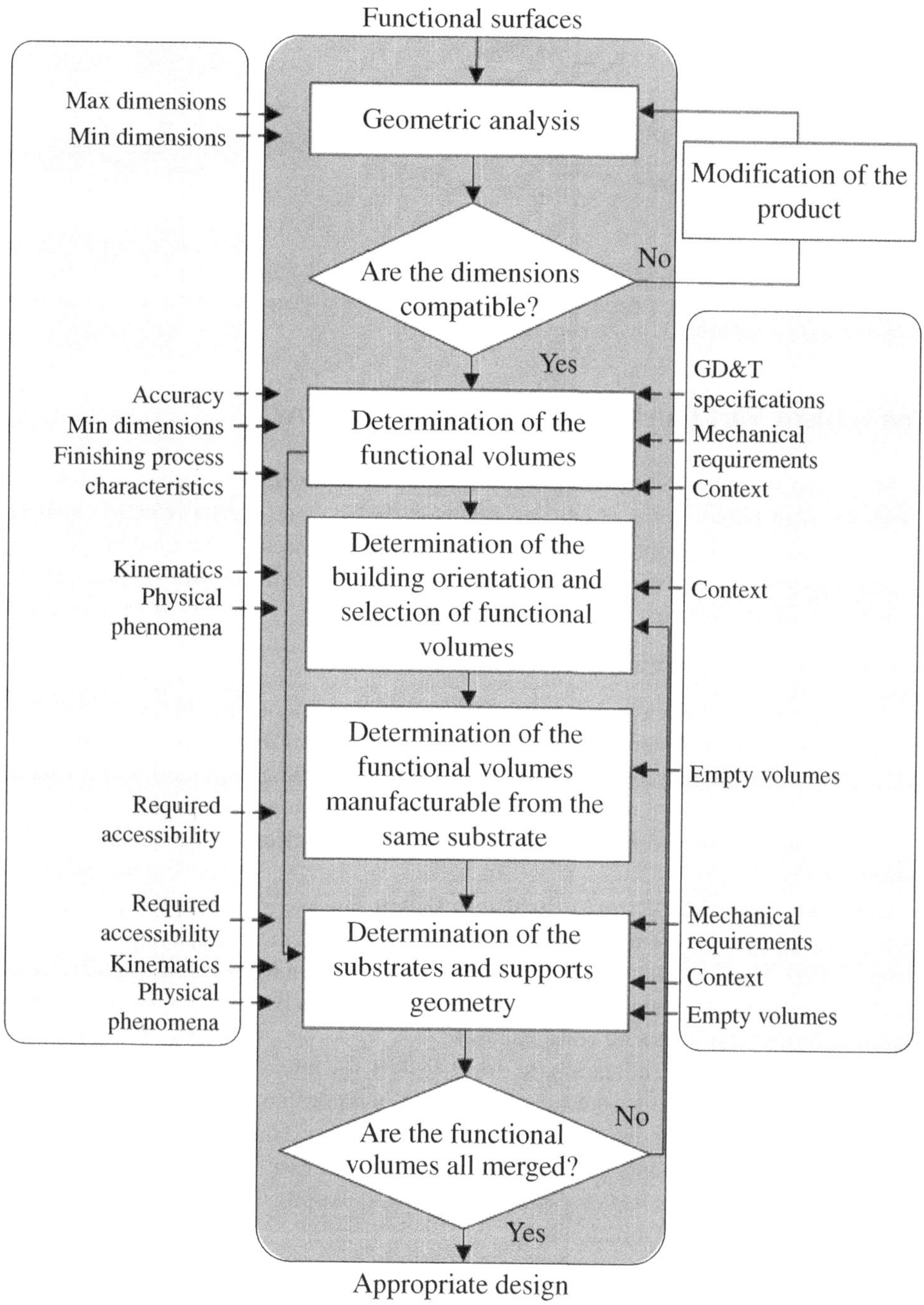

Figure 4.5 Functional surface and functional volume-oriented design method. *Source:* Adapted from [24].

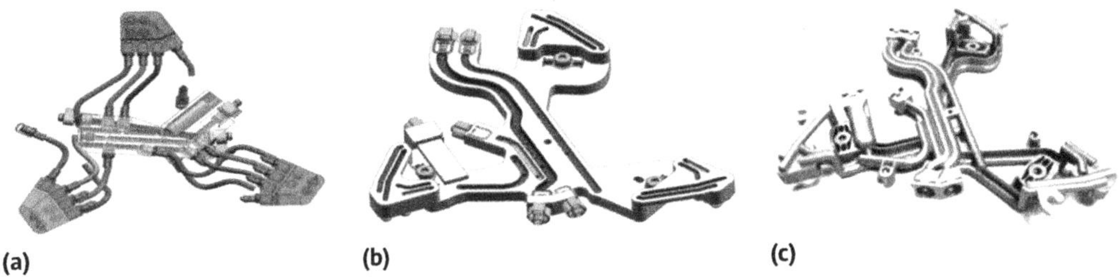

Figure 4.6 Three designs of a manifold in the semiconductor industry. (a) conventional design made of PEEK with hoses (110 g), (b) monolithic design milled in TiGr5 (200 g), and (c) optimized design printed in TiGr5 (110 g). *Source*: Adapted from [27]/with permission of Elsevier.

4.3 The Holistic Part Consolidation Design Workflow

A holistic PC design workflow is introduced in Figure 4.7. The workflow entails an assembly redesign scenario with initial inputs of existing assembly design (mostly refers to CAD model) and design requirements (e.g., boundary conditions and customer needs). The final output is a consolidated design with AM considerations. The workflow has four main stages: functional analysis, part candidate selection, fusion of parts, and refinements.

The main activities of each stage are as follows:

- **Functional analysis**. Functional analysis (FA) is to understand the design intent and uncover function structure and product layout, thereby establishing function-part mappings. FA has been intensively studied in the function modeling community and useful techniques can be found in review papers [28, 29]. The outputs from FA are the function structure and product architecture that functionally describe the design intent of the given assembly.
- **Part candidate selection**. A part candidate selection process is to find the part candidates for consolidation out of a given system according to a set of candidacy rules (e.g., no relative motion). These candidacy rules must examine the technical feasibility, quality, and cost margin of PC. The candidacy rules will be introduced in Section 4.4. The outputs of this module are the part candidates for consolidation.
- **Fusion of parts**. Once the candidacy parts are identified, the following task is to consolidate these parts into one functionally equivalent part. In this module, the critical tasks to be solved include 1) how to implement design for function paradigm to maximize performance and 2) how to construct the preliminary embodiment design that fulfills all functional requirements while taking consideration of AM characteristics. The output of this module is a consolidated IDS that can deliver required functions.
- **Refinement**. The preliminary embodiment design completes the transition from an assembly design to one single consolidated counterpart. Efforts are still in need to explore the chance of employing other AM potentials (e.g., topology optimization) for better performance and secure the manufacturability of the consolidated part. Therefore, other DfAM techniques introduced in other chapters can be integrated into this module, such as topology optimization, lattice design, and design rules.

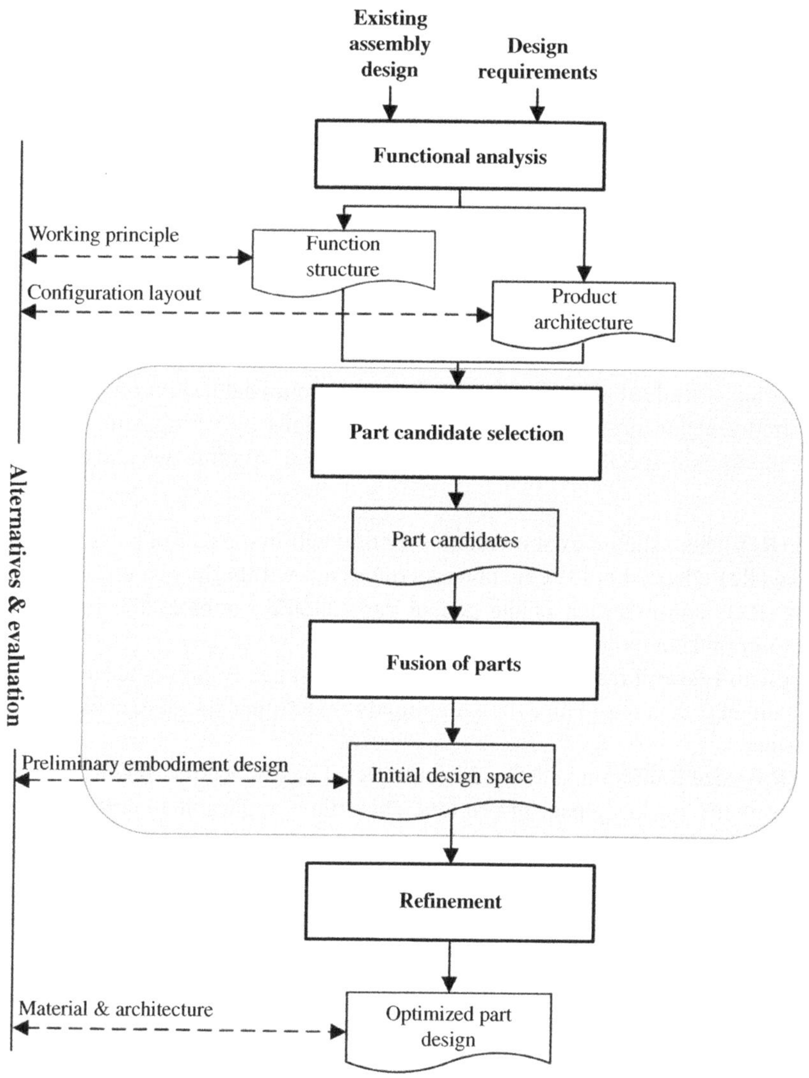

Figure 4.7 The holistic part consolidation workflow. *Source:* Adapted from [27] with revision.

Through this workflow, other design alternatives are also encouraged via integration with innovative design techniques such as brainstorming and TRIZ. By changing the working principles, configuration layout, initial design space, material, or architecture, various design solutions can be found. It should be noted that the earlier stage the alternative is located at, the larger difference in appearance will be between the original assembly design and the consolidated design.

Since FA techniques have been introduced in many literatures [28, 29] and part refinement techniques are covered in other chapters, the main focus of this chapter will be on the part candidate selection and fusion of parts as highlighted in Figure 4.7.

4.4 Part Candidate Selection

Part candidate selection solves the problem of "which parts are suitable for part consolidation." To answer this question, a set of candidacy rules that filter feasible parts are needed. The following question is "how to group part candidates in subgroups" because it is not essential to consolidate all feasible parts as one piece (e.g., modularization needs and printer volume limitation). To answer this question, a set of principles are introduced to rank the priority of being consolidated for each component.

4.4.1 Candidacy Rules

According to the pioneering work [30] and the ruleset r1–r9 as summarized in Section 4.2.1, the following rules are introduced for supporting part selection decisions by taking into account AM-enabled design freedoms, AM-specific process constraints, and cost. Assume two parts as Part A and Part B.

1. **Relative motion (R1)**: There should exist no relative motion within the pair of parts.
2. **Material variance (R2)**: There should be no material difference within the pair of parts.
3. **Assembly access (R3)**: Consolidation of the pair of parts should not block the assembly access of any other internal components.
4. **High-quality electronic/standard device (R4)**: Neither Part A nor B should be a low-cost standardized component (e.g., a round pipe) or a high-quality electronic/standard device (e.g., precision ball bearing).
5. **Size limitation (R5)**: Size limitation is defined by a triplet (Length, Width, Height), which normally is the volumetric working space of printers. This rule is applicable to both a single component and properly assembled component groups.
6. **Material availability (R6)**: Not all the engineering materials are currently available for AM processes.
7. **Maintenance frequency difference (R7)**: There should be no significant maintenance frequency difference within the pair of parts.

Rules R1–R4 are for quality concerns. Although AM-PC applications such as functionally graded material, articulated mechanism, and embedded foreign components have been proven feasible, these applications are highly case-specific and only serve as a proof of concept. Resulted quality of such cases is normally unsatisfactory. Rule R4 is of high significance because rare industrial applications of printed electronics in high quality have been reported. Massive produced standardized components (e.g., tubes) are still financially favored, especially those with high precision requirements (e.g., ball bearings). Rules R5–R6 are for concerns of AM general constraints. Some may argue why specific constraints of AM are not considered in this ruleset. The main reasons are as follows. The specific constraints of AM are highly form dependent. Feasibility analysis (e.g., overhanging) based on the current assembly architecture is meaningless because consolidated design unnecessarily adopts the same shape and geometry. Therefore, specific constraints of AM should be postponed to the timeline when details of the consolidated design are enriched (i.e., refinement phase of the holistic workflow), thereby speeding up the whole decision process. Rule R7 is for maintenance and disassembly concerns. Consolidation of parts having similar maintenance frequency saves maintenance cost and leads to simpler retirement treatment.

4.4.2 Priority Principles of Part Consolidation

This set of fundamental principles is for the concern of preserving functional proximity, improving performance, and lowering cost under the premise that all feasibility rules are conformed. Functional proximity indicates the closeness and interaction intensity within a component pair or a component group. Therefore, functional proximity can be used as the criterion of prioritizing consolidating sequence if consolidating all candidates is infeasible. Functional proximity can be described in three scales: module, community, and pair-wise strength. Details are covered as follows. More details can be found in literature [31].

4.4.2.1 Principle of Modularization (Principle I)

Modularization is one of the critical concepts in product design [32–34], through which it is believed to reduce design complexity, reduce development cost, increase configuration flexibility, enable distributive design, and control system failure propagation [35]. A module can be defined as "a relatively independent chunk of a system that is loosely coupled to the rest of the system" [36]. This relative independence amongst modules helps to reduce the amount of information to be handled in each module, which enormously lowers the design expertise requirements at each level [37]. Therefore, it is necessary to incorporate the need for modularization in the process of grouping part candidates.

In contrast with the way of partitioning a system into discrete modules and creating interfaces in between, PC is a means of eliminating interfaces between components to achieve an integrated design that outperforms modular design in system integrity and simplicity. Trade-offs must be made between modular design and integrated design (i.e., PC). PC design essentially studies the "local" pair-wise relation between component pairs while modular design offers a "global" view of a chunk of components. On this basis, PC design can be seen as the local modifier of the globally optimized modular division result. Therefore, benefits from both PC design and modular design can be enjoyed. Hereby, the first principle of PC design is phrased as:

Principle I: *Part consolidation candidates should be primarily conformal to modules and inter-module parts are discouraged from consolidation.*

This principle is generally applicable especially for highly complex systems. In this principle, a module is treated as a hard constraint for PC; therefore, how to reasonably define modules is of significance. This topic has been well-examined in prior studies [33, 38, 39]. The basic requirement to be fulfilled by any modularization method in the PC context is that it should directly break a given system into the lowest-level modules while securing the maximum modularity. In other words, any further subdivision of the modularization result does not contribute to a higher modularity. Modularity is a measure of the degree of dependency and similarity of components within modules and that of independency and dissimilarity of components across modules [35]. In practice, the computational method of modularization to maximize modularity is not always suggested for extremely complex systems (e.g., a vehicle or an airplane) due to cost efficiency concerns. Using module information (i.e., subsystems) that already exists in a highly modularized product as the starting point for grouping PC candidates is more favored.

4.4.2.2 Principle of Community (Principle II)

The concept of community was initially introduced in the social or biological discipline to describe the phenomenon that intra-community connections are intensive while inter-community interactions are sparse. This phenomenon is defined as community characteristics and it is also observed in mechanical systems [40]. Community and module are essentially the same in that both investigate the internal structure properties of a system. The desire to introduce the concept of community rather than sub-module is to avoid confusion with Principle I. As stated in the principle of modularization, there is no need to break modules into sub-modules if the global maximum of modularity is reached in the first place. However, community characteristics inside a module should not be ignored because functional and physical dependency amongst intra-community components is higher than that of inter-community ones. To preserve such dependency, the priority of consolidation should be given to intra-community components. Therefore, the second principle of PC design is phrased as:

Principle II: *Intra-community component groups should have a higher priority of being considered for consolidation than inter-community ones if consolidating all components is infeasible.*

This principle offers a reasonable solution of partitioning a component group into smaller groups to satisfy dimensional constraints (e.g., R5). Different from Principle I, this principle only regulates the priority of being consolidated but it does not forbid consolidation of components located in different communities if feasibility rules are conformed. The underlying reason is that information to be processed in a module scale (compared to a product scale) has already been reduced to a manageable level within a design team. Regarding a community as a module (i.e., hard constraint) may hinder the opportunities of improving system integrity and reducing cost. Therefore, a community should be treated as a weak constraint.

4.4.2.3 Principle of Strength (Principle III)

In contrast with Principles I and II that focus on groups of parts (i.e., more than three parts), this principle investigates pair-wise relations of components. The concept of strength is introduced to characterize the level of functional and physical dependency between a pair of components. Therefore, the third principle is phrased as:

Principle III: *The component pair with higher functional and physical interaction strength should have a higher priority of being considered for consolidation than the adjacent pair if consolidating all components is infeasible.*

One exemplified case is shown in Figure 4.8. The tight coupling relation between gear and shaft is to transmit large radial force while the axial force between the gear and the bearing is relatively small. Therefore, the interaction strength of the former is higher. Consolidation of shaft and gear has higher priority than that of gear and bearing.

4.4.2.4 Principle of Minimum Part Count (Principle IV)

Prior studies have revealed that part count reduction leads to a significant decrease in assembly cost [41], and an even larger saving in supply chain management cost [12]. Apart from cost advantages, combining the principle of "minimum part count" and TRIZ [15] can enable designers to think more creatively [42]. Hereby, the fourth principle is phrased as:

Principle IV: *Conformal to all rules and other principles, the design solution with the minimum part count should be favored at least in the screening stage.*

This principle implies that further breakdown of a feasible component group is unnecessary especially in the part filtering stage. Making a feasible part group as large possible in the part

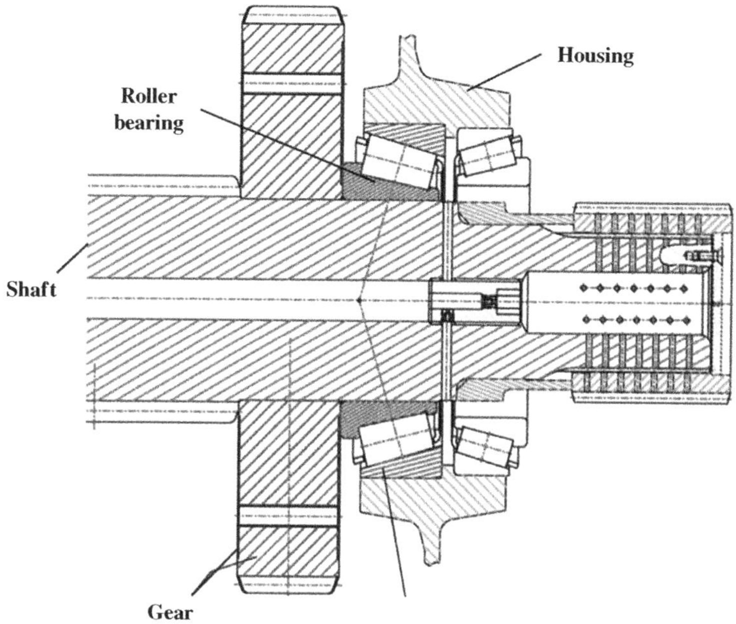

Figure 4.8 An example of illustrating functional and physical interaction strength.

filtering stage can enlarge the solution space in the refinement stage (see the holistic workflow), thereby effectively avoiding prematured solutions. One extreme case of prematured solutions is to break feasible groups into individual components, which is definitely a feasible design solution (i.e., the original design) but not the optimal.

4.4.2.5 Principle of Performance (Principle V)

All the aforementioned rules and principles focus on securing feasibility and lowering cost. However, one of the driving forces to use AM has not been considered. This driving force is the capability of improving performance. Therefore, the fifth principle is phrased as:

Principle V: *If consolidation leads to a big leap in performance improvement, all the other rules and principles can be neglected.*

This principle grants designers certain degrees of freedom to test more innovative solutions regardless of cost and other restrictions. However, to realize such disruptive innovation, designers' creativity is the deterministic factor. This principle is associated with the generation of alternative solutions in the holistic PC design workflow.

4.4.3 Part Candidate Filtering Process

The part candidate filtering process with integrated candidacy rules and principles are shown in Figure 4.9. This process needs initial inputs of functional structure and product architecture (CAD models) to establish the functional and physical interaction relations amongst components. Through stepwise instructions, designers should be able to efficiently find the grouping solution with the minimum part count.

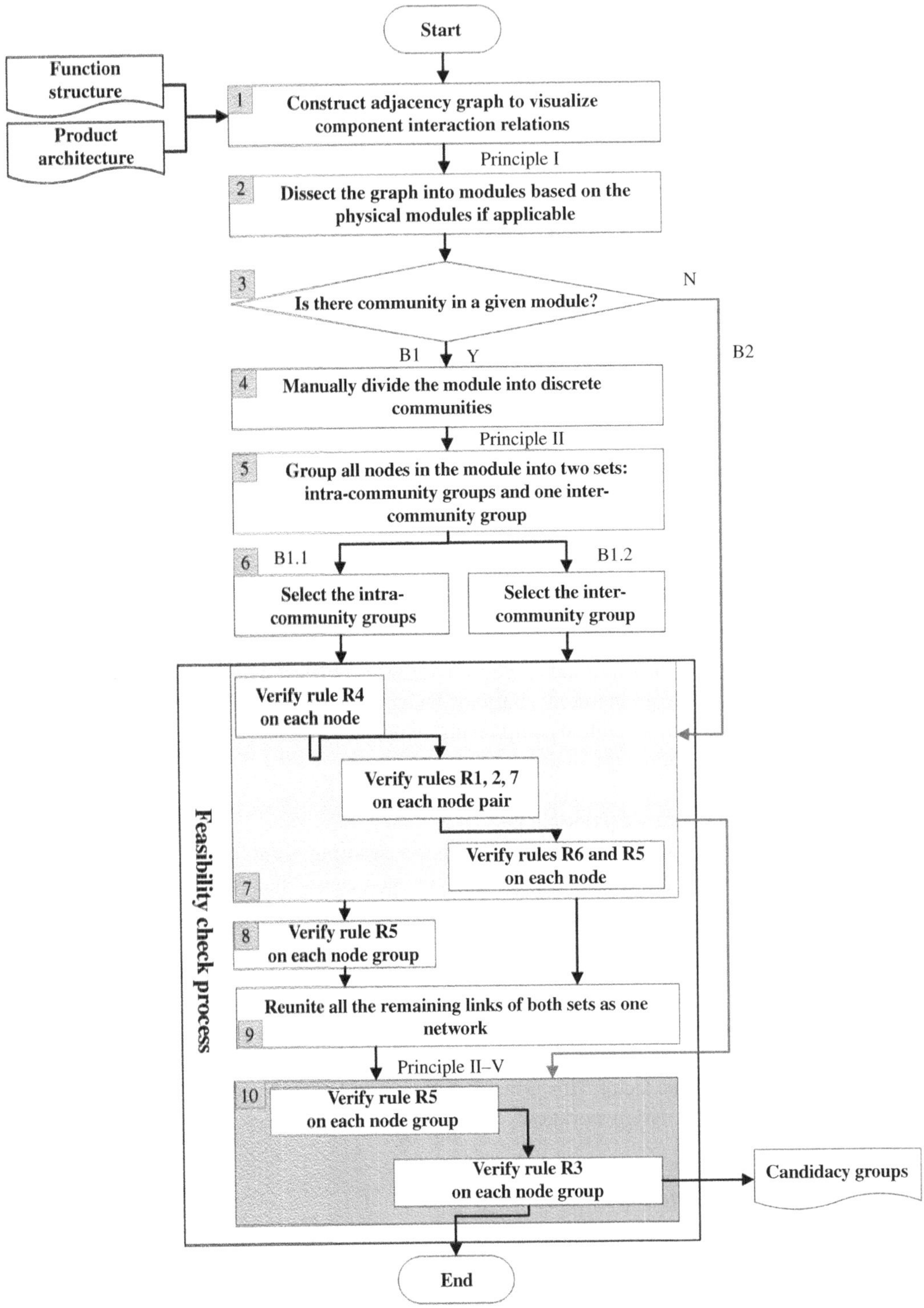

Figure 4.9 The proposed heuristic screening process.

Detailed steps are presented as follows.

- Step 1. An adjacency graph should be constructed to visualize the functional and physical relations amongst all components of the original assembly. If there exists functional or physical dependency between a component pair, a link is established; otherwise, no link is drawn.
- Step 2. The designer should decide whether to dissect the given design into modules (Principle I). This step is highly dependent on the scale of the design input. If the input is only an assembly (with a few components) or a module (e.g., power supply unit of a computer), this step is negotiable. However, if the input is highly complex and involves multiple disciplines, module division is recommended.
- Step 3. For a given module, a decision must be made whether there exist communities inside the module. If yes, continue the main branch (B1); otherwise, go to the second branch (B2) marked in red.
- Step 4. Divide the module into reasonable communities (Principle II).
- Step 5. Group all components of the module into two sets: intra-community groups and an inter-community group. The formal one includes all disconnected communities while the latter one joins all inter-community component pairs.
- Step 6. Two parallel subbranches (B1.1 and B1.2) are derived in accordance with the two sets in Step 5. The main reason is that feasibility verification process of these two parallel streams is different because of Principle II and efficiency concerns. Details are reported in Step 8.
- Step 7. This is a shared module for all three branches (B1.1, B1.2, and B2). The purpose is to sequentially examine component and component pairs in terms of candidacy rules R4, R1, R2, R7, R6, and R5.
- Step 8. This is a unique step for branch B1.1 because there is a need to examine whether the component group in a module is conformal to R5 (size limitation) because of the expanded group size.
- Step 9. Branches 1.1 and 1.2 are recombined as one graph.
- Step 10. This is a shared module for branches B1 and B2. The purpose is to partition the newly combined graph (for branch B1) or the remaining connected graph (for B2) into smaller feasible groups against the rules R5 and R3. It should be advised that iterations are needed between the sub-steps of checking R5 and R3. In the end, candidacy groups satisfying all rules and principles are sought.

In the feasibility check process, there are a few aspects to be clarified. Firstly, AM process-specific constraints are necessary for verifying rules R5 and R6 in terms of available materials and allowable building volume. Secondly, all steps of this filtering process can be manually implemented. Heuristic modularization/community partition can be done by either reserving the existing physical modules (i.e., functional chunks such as a battery unit) or updating the modular design via functional flow-based approaches [34]. Lastly, the proposed heuristic rules and principles help to effectively down-select the potential combinations and generate the minimum grouping solution(s).

4.4.4 Throttle Pedal Redesign Example

A throttle pedal redesign example is presented to demonstrate how the filtering process works. The original design is retrieved from an open-source CAD database [32]. As shown in Figure 4.10a, the

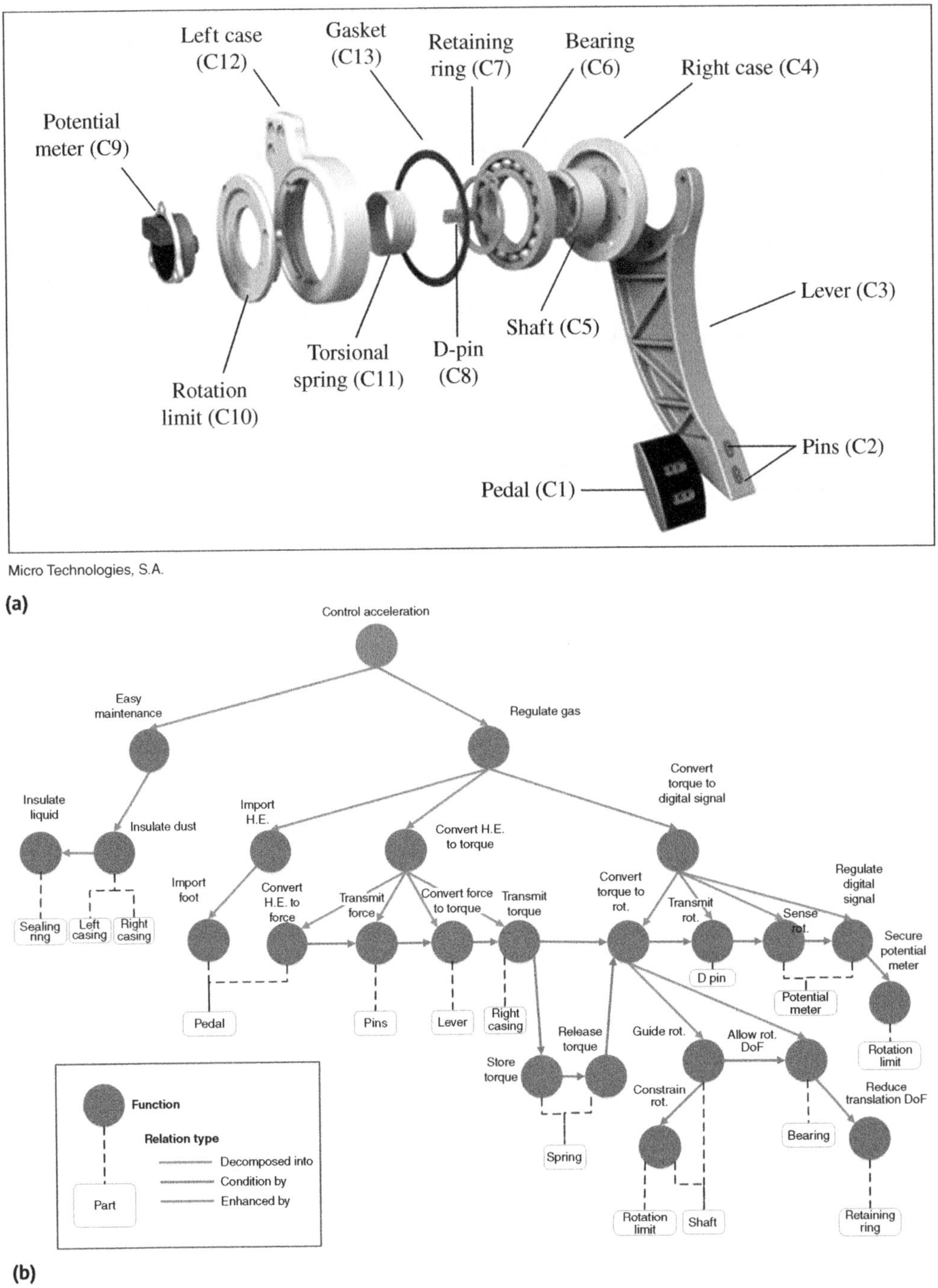

Figure 4.10 Function structure and product architecture of the original assembly design. (a) Physical decomposition view and (b) function-part mapping view.

product is comprised of 13 components without counting washers and fasteners. All parts except the gasket and potential meter are made of steel. Among these parts, only the bearing and potential meters are standard components. This product functions as a device to convert human energy to a digital signal so that drivers can control the gas emission and thereafter accelerating rate. It is also required to fulfill the requirements of providing certain degrees of resistance when the driver steps on the pedal, restraining the angle of rotation within 40 degrees, allowing the pedal to bounce back when it is released, and sustaining a maximum load of 250 lbs. Through the iterative steps of function decomposition and product breakdown, function-part mappings can be easily established as shown in Figure 4.10b. It can be observed that there exist 1-to-n (e.g., constrain rotation via parts—shaft and rotation limit) and n-to-1 (e.g., store torque and release torque via torsional spring) function-part mappings.

Some basic physical attributes of each component in the throttle pedal assembly are summarized in Table 4.1 including dimensions, material, maintenance frequency (MainFreq), and whether it is a standard part (IsStandard). Meanwhile, the relative motion information between each connected part pair is summarized in Table 4.2. These information can be easily retrieved from the CAD model and design files to form basic understanding of functional and physical relationships between each pair of components. The selected printer is Renishaw AM250 whose maximum building volume is (L250, W250, H300). Available materials for AM are limited to steel and plastic only for test purposes.

The detailed steps of implementing the part candidate filtering process (see Figure 4.9) are as follows:

- Step 1. An adjacency graph is constructed based on the functional and physical interactions between each pair of parts in the throttle pedal assembly. As shown in Figure 4.10b, functional interactions between parts can be in the form of transmitting energy (force or torque), signal, or material. Physical interactions between parts can be attachment (e.g., fasteners or weld alike), positioning, and containment (e.g., parts sharing the same housing). The interaction strength between each pair of parts can be graded according to the criteria as suggested in Table 4.3. The resulted adjacency graph is shown in Figure 4.11.
- Step 2. The throttle pedal is a simple assembly and well self-contained so that there is no need to dissect it into modules (Principle I).
- Step 3. According to the adjacency graph obtained in Step 1, the graph shows clusters of parts. For example, parts C1, C2, and C3 are firmly connected as one cluster while parts C4, C8, and C12 are closely connected. As such, we need to identify those clusters (i.e., communities) where internal parts have higher priority of being consolidated than other outside ones.
- Step 4. Dividing the adjacency graph into reasonable communities (Principle II). By breaking the weak links (i.e., weight of 0.2) and minimizing the connections between communities, we can get four communities in total as shown in Figure 4.12a. Nodes in the same color belong to the same cluster.
- Step 5. Grouping all components of the adjacency graph into two sets as intra-community part groups and an inter-community part group as shown in Figure 4.12b and c, respectively.
- Step 6. Choosing the intra-community part groups and applying candidacy rules for each part and each pair of parts for each group in terms of candidacy rules R4, R1, R2, R7, R6, and R5. For example, the link between C1 and C2 is disconnected because of maintenance frequency difference. Parts C5 and C6 are disconnected because of relative motion. The resulted graph is shown in Figure 4.12d. Repeat the same process for the inter-community part group and the resulted graph is shown in Figure 4.12e.

Table 4.1 Component list and physical attributes.

		Pedal	Pins	Lever	Right case	Shaft	Bearing	Ret. ring	D-pin	Potl. meter	Rot. limit	Spring	Left case	Gasket
#		C1	C2	C3	C4	C5	C6	C7	C8	C9	C10	C11	C12	C13
Volume	**L**	50	11	217	74	40	68	52	28	40	63	38	107	74 74
	W	37	7	74	74	40	68	52	7.5	40	63	34	74	0.5
	H	10	6.3	22	12.5	23	9	1.6	5.5	23	9	20	18	
Material		Steel	Steel	Steel	Steel	Steel	Steel	Steel	Steel	Plastic	Steel	Steel	Steel	Rubber
MaintFreq		H	L	L	L	L	L	L	L	L	L	L	L	L
IsStandard		0	0	0	0	0	0	0	0	1	0	0	0	0

Table 4.2 Relative motion between each connected part pairs.

	C1	C2	C3	C4	C5	C6	C7	C8	C9	C10	C11	C12	C13
C1													
C2													
C3													
C4												1	
C5						1				1	1		
C6							1					1	
C7													
C8													
C9													
C10											1		
C11													
C12													1
C13													

Table 4.3 Scoring criteria based on interaction strength between components.

#	Strength	Value	Description		Examples
			Physical interaction	Functional interaction	
1	Very strong	1	Firm connection	Exchange of large force or torque	Tightly bolted connection
2	Strong	0.8	Medium connection	Exchange of large force or torque	Multiple-bar mechanism
3	Medium strong	0.6	Firm connection	Exchange of medium force or torque	Non-laden welded frame
4	Medium	0.5	Medium connection	Exchange of medium force or torque; Exchange of energy	Cable and battery
5	Medium weak	0.4	Loose connection	Exchange of low force or torque	Left and right casing
6	Weak	0.2	No direct connection	Exchange of signal or material	Positioning/channel
7	No	0	No	No	

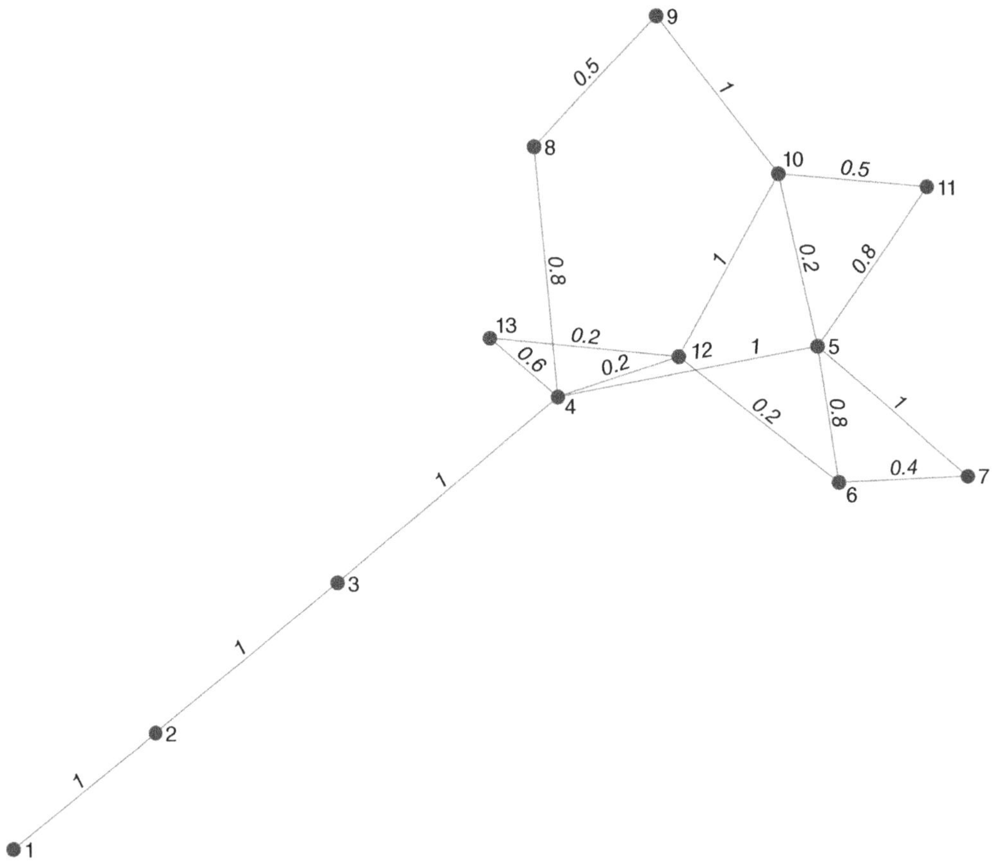

Figure 4.11 The adjacency graph of the throttle pedal. Each node represents one part. The link between two nodes indicates the existence of functional or physical interactions in between. The weight is determined by the interaction strength (e.g., the magnitude of forces).

- Step 7. Assembling the remaining intra- and inter-community part groups as one united group. The resulted graph is shown in Figure 4.12f.
- Step 8. Applying the size limitation constraints for the united group. If size constraints are violated, disconnect the links in the sequence of inter-community links first and then weak links within each intra-community group. No size limitation is violated for the whole united graph in this example. The resulted graph is shown in Figure 4.12g.
- Step 9. Applying the assembly access rule (R3) to the united graph. It is found that consolidation of parts C5 and C7 will block the assembly access of C6, which cannot be consolidated with the rest; therefore, the link between C5 and C7 is disconnected. The final selected part candidates for consolidation are shown in Figure 4.12h. There are two separate part groups identified: (C2, C3, C4, C5, C8) and (C10, C12). Corresponding physical parts in each group are shown in Figure 4.13.

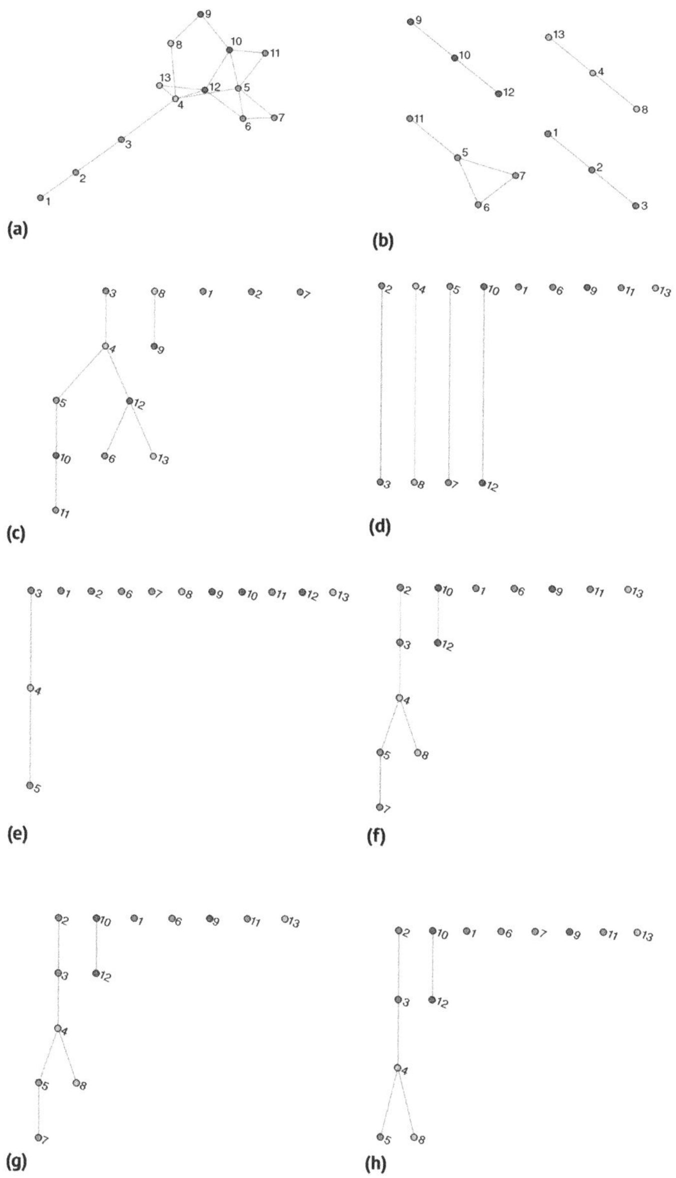

Figure 4.12 Part candidate selection process for the gas pedal assembly. (a) Community division results after applying Principle II, (b) intra-community part groups, (c) inter-community part groups, (d) intra-community part groups after applying rules R4,1, 2, 7, 6, and 5, (e) inter-community part groups after applying rules R4,1, 2, 7, 6, and 5, (f) assemble inter- and intra-part groups as one united group, (g) united group after applying rules R5, and (h) united group after applying rule R3. Node clusters (1, 2, 3), (4, 8, 13), (9, 10, 12), and (5, 6, 7, 11) are colored in red, green, blue, and brown, respectively. (Please refer to the online version for the color image.)

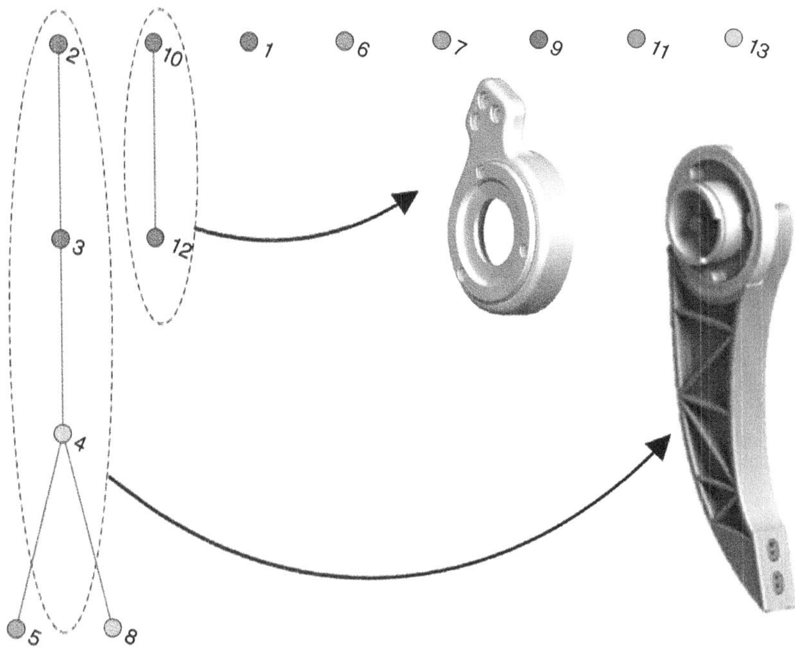

Micro Technologies, S.A.

Figure 4.13 Identified part groups for consolidation.

4.5 Fusion of Parts

Fusion of parts is not direct elimination of the assembly interfaces; more importantly, it aims at creating a functionally equivalent consolidated design by rethinking about the architecture. To achieve this, identification of the essential features that should remain in the consolidated design and avoidance of obsolete features that were designed for conventional processes are the top tasks. These essential features are defined as functional entities. By rejoining the discrete functional entities as one piece will contribute to the IDS. To aid the transition process, a functional entity-based method is introduced to find the IDS. The consolidated IDS can be further optimized in the refinement stage as discussed in the holistic PC workflow.

4.5.1 Functional Entities

Functional entities include three types of features: functional interface, functional volume, and initial design space. The definitions are introduced in the following subsections.

4.5.1.1 Functional Interface

A functional interface (FI) is the interface by which the flow of a function is transformed or bypassed. For example, a torsional spring can transform human energy as potential energy, and the FI is the flat end that interacts with human hands. The other example of an electric cable is to conduct electricity without transforming anything, and the FI is the plug pin. A FI can be in the form of a point, a line, or a surface depending on its specific functions. Representative examples of FIs include the tip (point), cutting flute (line), and shank (surface) of a ball end mill (surface). Since surface contact is the most popular in engineering applications, FI is also named as functional surface (FS). Without special notations, FI and FS are interchangeable in this chapter. There are three types of FIs that can be identified according to the interaction types.

- **Part-environment FI**: This type of FI serves as the interface between the artifact and external environment to fulfill functions related to water/gas conveyance (e.g., manifold), aerodynamics (e.g., windshield glass), or support (e.g., bottom surface of a mug), etc. It is also consistent with the environment-centered view of function as the effect on environment [43].
- **Part-user FI**: It mainly provides medium to import or transform energy (e.g., buttons), material (e.g., fingerprint), or signal (e.g., aesthetics or voice detector) from human. This type of FI is consistent with user-centered view of function as design intent [44].
- **Part-part FI**: It may not exist in an assembly-free design/product (e.g., a product consisting of a single part). However, most systems of interest are a part of a large system; therefore, having assembly interfaces with external systems is necessary. Such FI is to fulfill functions related to position, couple, or allow DoF (degrees of freedom).

An FI can be modeled as a set of key surfaces with function, shape, dimensions, and attributive information as represented in Equation 1 as:

$$\mathrm{FI} = \left\{\left(S_f, Fr\right) \middle| S_f(\mathrm{Brep}, \mathrm{Post}, \mathrm{Norm}, \mathrm{Attr})\right\} \tag{4.1}$$

Where the dual-tuple $\left(S_f, Fr\right)$ represents the surfaces and corresponding function requirements fulfilled by the FI. Each S_f can be described by Brep (boundary representation), Post (position), Norm (normal direction), and Attr (attributes). Norm indicates the active side of a surface. Attr should include material, color, and texture information to enable full descriptive capability of shape, material, and color complexity in AM context. Current CAD systems show strength in modeling capability of Brep, Post, and Norm but lacks support for describing Attr.

4.5.1.2 Functional Volume

A FV is defined as the enclosed space that is formed by thickening the FI so that the desired function can be fulfilled. The thickening direction is normal to the surface, and the thickness (T) of FV is determined by the minimal thickness (t) corresponding to local mechanical requirements and residual material (α) for the finishing process. The relation in between is as follows.

$$T = t + \alpha \tag{4.2}$$

$$\alpha \geq \beta \tag{4.3}$$

Where β is the dimensional accuracy of the selected AM process. Equation 2 holds regardless of whether the surface is an internal or external surface. As shown in Figure 4.14, Surfaces 1 and 2 are the desired surface and as-built surface, respectively. Surface 3 defines the minimal thickness as required.

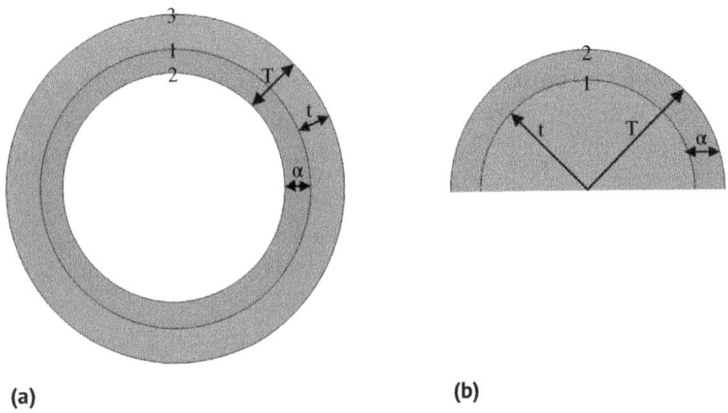

Figure 4.14 Relation between thickness, minimal thickness, and residual material. (a) Internal surface and (b) external surface.

4.5.1.3 Initial Design Space

IDS is described as the physical body that integrates all discrete FVs while being subject to system boundaries (SBs). Based on this definition, the relations between the FI, FV, and IDS can be expressed as:

$$\mathrm{FI} \subseteq \partial\Omega_{\mathrm{FV}} \tag{4.4}$$

$$\mathrm{FV} \in \mathrm{IDS} \tag{4.5}$$

Where $\partial\Omega_{\mathrm{FV}}$ represents the boundary of an FV that belongs to the IDS. The selection of IDS is very important because the maximum dimensions defined by IDS affect the morphology of the final design solution. An example is shown in Figure 4.15. The same pressure load and boundary conditions are applied to three cantilever beams with various IDS (varying in height dimensions). Different desired volume fractions are applied to make sure of the same weight constraint for each scenario. The optimized results indicate the drastic difference of final topologies and it is observed that the larger IDS leads to a smaller displacement. However, determination of a proper IDS in the first round is challenging. As such, iterative design-validation cycles are necessary.

4.5.2 Functional Entity-based Design Method for Fusion of Parts

The functional entity-based design flow is shown in Figure 4.16. The initial inputs for the process include a geometric model and boundary conditions of the assembly design. The final output is a single IDS satisfying all functional requirements and boundary conditions. This generated IDS is in preparation for the downstream refinement process such as topology optimization; therefore, two points should be advised. 1) BO is determined at the embodiment design stage and FIs are modified accordingly to secure manufacturability. 2) FVs generated in the process are equivalent to the non-design space (in topology optimization) that material should be reserved while the subtraction of IDS and FVs is the design space where optimal connecting architecture between FVs can be computationally found. The detailed steps are illustrated as follows.

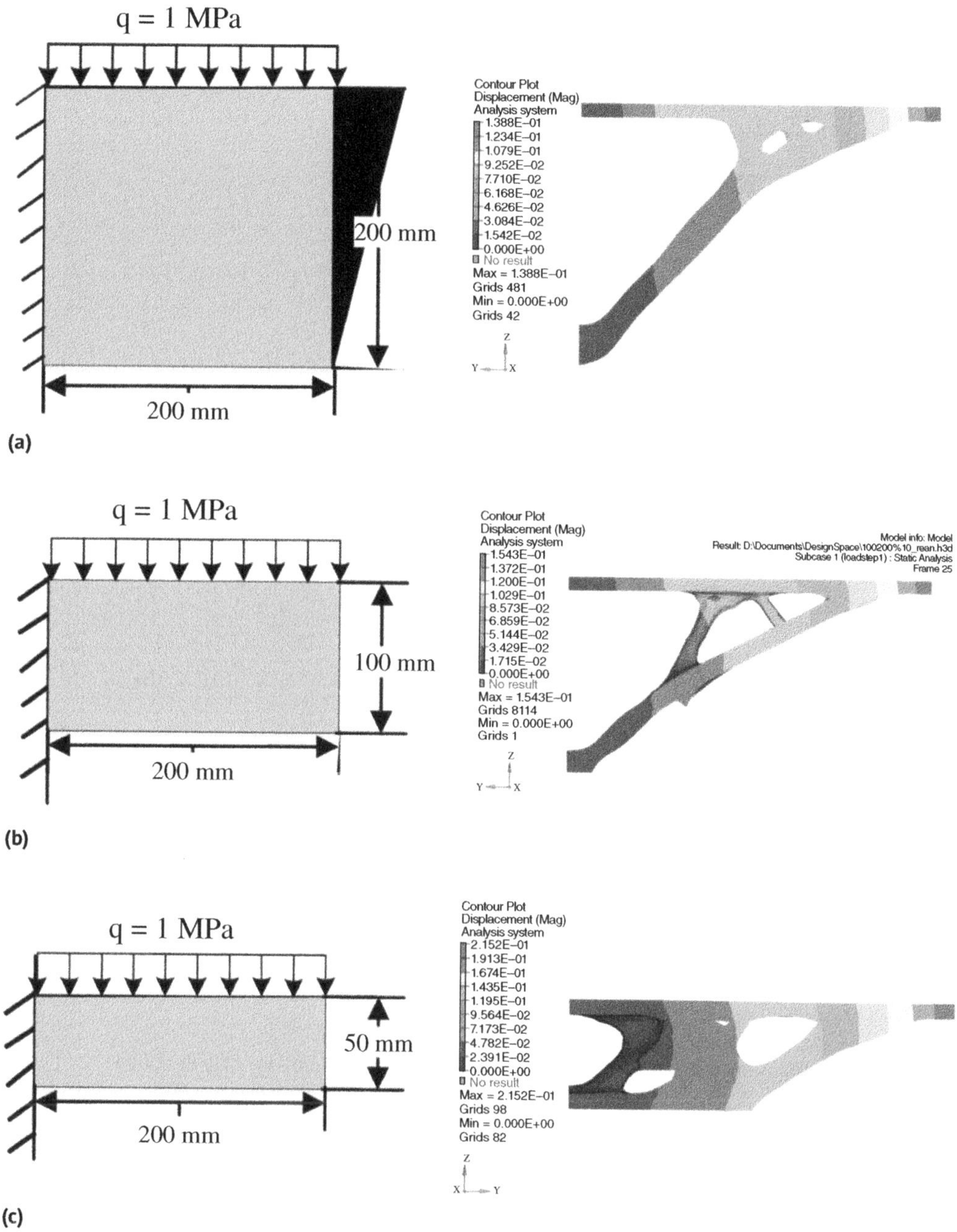

Figure 4.15 The effects of IDS on final design solution. (a) Volume fraction 5%, (b) volume fraction 10%, and (c) volume fraction 20%. *Source*: [45]/with permission of Yunlong Tang.

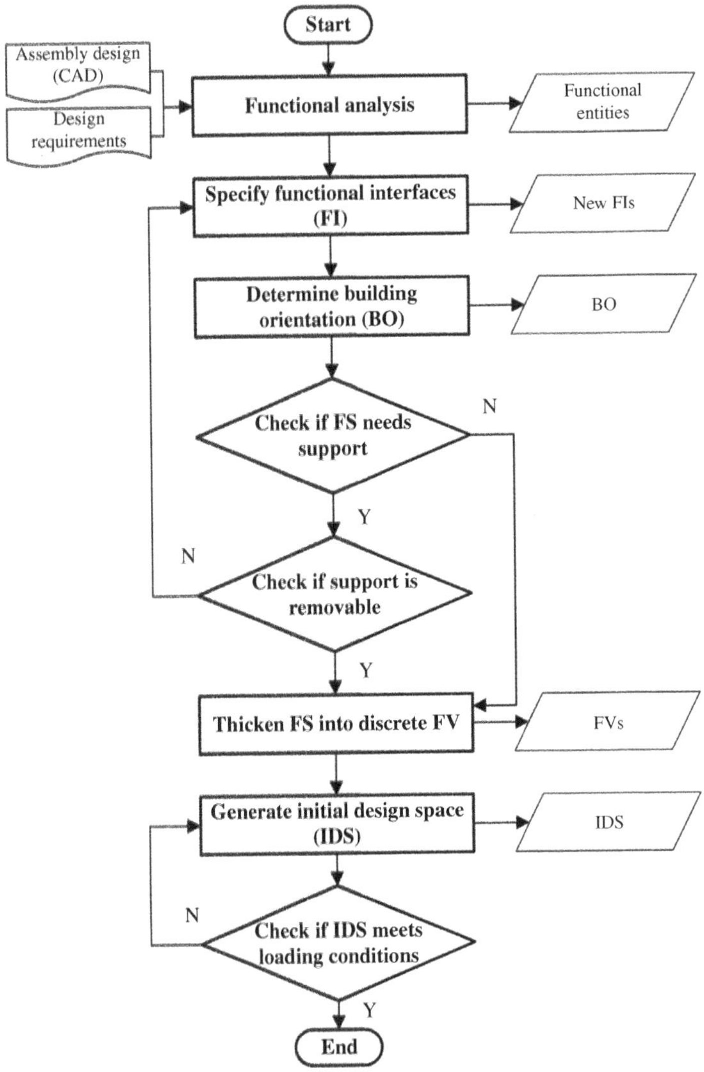

Figure 4.16 The proposed embodiment design flow making transition from assembly design to consolidated design.

As it is common in a redesign scenario that a CAD model does not provide enough context of design specifications and functional information, a FA process is essential for uncovering the design intent, identifying SB, and establishing function-FI mappings. The extracted FIs and SBs serve as cornerstones of defining the morphology of the consolidated part. However, these FIs are not suitable for direct reuse in the consolidated design. There are two reasons: 1) it may consist of function-irrelevant features such as obsolete machining features (e.g., undercut), and 2) extracted FIs may compromise in performance because of process constraint considerations. Therefore, extracted FIs (except part-part FIs) should be replaced by new ones that are synthesized by advocating design for performance. Next, the optimal BO is determined to secure successful printing of these FIs instead of considering the whole part geometry. If it is impossible for all FIs to be support-free, the most critical FIs (e.g., difficult-to-access internal channels) should be ranked with higher priority. On the determination of the BO, two successive steps are taken to examine whether certain FIs need modifications to avoid support structure. Ideally, FIs should be self-supporting or at least allowing easy access of post-processing operations (e.g., machining or sandblasting). The iterative loops repeat until all FIs comply with manufacturability constraints. Then, all FIs are thickened into FVs and the thickness is carefully assigned. Afterwards, an IDS is constructed to encapsulate all the FVs with respect to the SB constraints. In the end, an iterative process is taken to refine the IDS by implementing the cycle of design and simulation. For the sake of better clarity, the detailed techniques of establishing function-FI mappings, synthesis of new FIs, determination of the optimal BO, and reconstruction of IDS are carefully elaborated in the following subsections.

4.5.2.1 Establishing Function-FI Mappings

Different from the common practice of FA to establish function-part mappings, fusion of parts requires function-FI mappings. One feasible solution is to associate a function with FIs in two levels: function-part mapping and part-FI breakdown. The basic principle is depicted in Figure 4.17. The first level is to establish function-part mappings through iterative function decomposition and product breakdown. For the completeness of comprehension, steps of the part-level FA are as follows. The initial inputs for this level of FA include the whole system of interest (rather than the part candidates) and design requirements.

1. Model the overall function of the system of interest with force-flow black box model. For example, the overall function of a gas pedal is to convert load to a digital signal.
2. Identify the major flow. For example, to transfer load for the pedal.
3. Identify a sequence of subfunctions by following Form Follows Form principle [17]. "Form" here refers to the type of flow (e.g., force flow). A sequence of a force flow of the pedal may be of the form: import force, transmit force, and convert force to potential energy, etc.
4. Repeat Steps 2 and 3 until the inputs and outputs of the overall functions are matched with those of subfunctions.
5. Aggregate subfunctions into a function structure by following the same type of flow.
6. Dissect the system of interest into parts and establish the product hierarchy.
7. Match subfunction(s) with part(s) to establish function-part mappings. If there exist the following mismatch situations, correct the function or add a new function: a) a function cannot find a matching part, or b) vice versa. However, it is advised that it is not always bijective mapping between function and part. There may exist n-to-1 or 1-to-n cases because of function integration and manufacturing constraints.

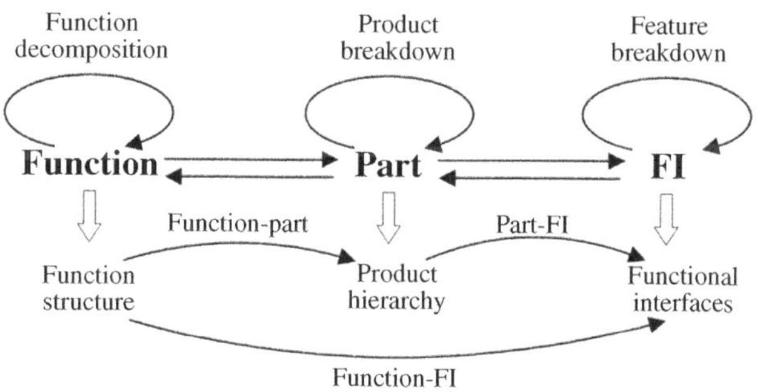

Figure 4.17 The dual-level FA method for establishing function-FI mappings.

Throughout the above steps, function basis [46] can be adopted to formalize the description of functions. The inputs for part-FI breakdown only include the candidacy parts and corresponding functions. Basic steps to be taken are summarized as follows:

8. Identify part-environment FI(s) of each part in accordance with its functions. This type of FIs normally interacts with external environment such as fluid, heat, gas, or solid support (e.g., desk). Moreover, the function delivered by this type of FIs is normally the primary concern, and the shape of FIs determines the basic layout of the part.
9. Identify part-user FIs of each part in accordance with functions related to "import human material" or ergonomics.
10. Identify part-part FIs of each part. Only those FIs that interact with parts outside of the candidacy group are reserved. In other words, internal mating interfaces are useless for consolidation.
11. Identify SBs that constrain the maximum dimensions of the whole part candidate group. Normally, SBs are in the form of voids and they do not have direct contact with external obstacles.
12. Perform design for manufacturing and assembly analysis for each part forms better understanding of its product architecture. This step is necessary to spot any uncovered function in the part-level function analysis. This is because features as a set of geometric entities help to record the design rationale, physical principles, and design rules in the design process [47]. More importantly, analysis of assembly features helps designers to uncover potential assembly access issues (e.g., tool access for fastening bolts), which is critical for defining the IDS.

This dual-level FA method helps to establish function-FI mappings and paves the way for identifying new opportunities for an alternative solution or optimizing the design parameters for better performance.

4.5.2.2 Synthesis of New FIs

Previously extracted FIs are not directly applicable for IDS generation. There are two main reasons. First, the final shape of an extracted FI may be a compound of multiple features for different purposes such as functionality, ease of assembly (e.g., chamfer), or manufacturability (e.g., round corners instead of sharp corners). As such, shape of the FI needs to be simplified to capture the real design intent. One example is shown in Figure 4.18a. The chamfer of the assembly hole is simplified but the topology remains unchanged because it is a part-part FI that matches with external parts. Second, existing FIs are a compromised result of desired performance and the process capability of CM methods. Because of the expanded processing capacity of AM, it is worthy of synthesizing a new FI with better performance. One good example of a new FI is shown in Figure 4.18b. The original straight channel is to comply with process constraints but the new FI is redesigned to be conformal to the contour of the core to maximize performance. Two possible strategies can be applied to find an alternative design solution: analogical design and computer-aided design synthesis [48]. Analogical design heavily depends on knowledge of existing artifacts or nature, which serve as sources of inspiration to stimulate design creativity. In contrast, computer-aided design synthesis employs the advancement of computational power and artificial intelligence to generate optimized geometric entities on the premise of well-defined objectives and boundary conditions. Representative design tools include shape optimization, FEA, CFD, generative design tools [49], etc.

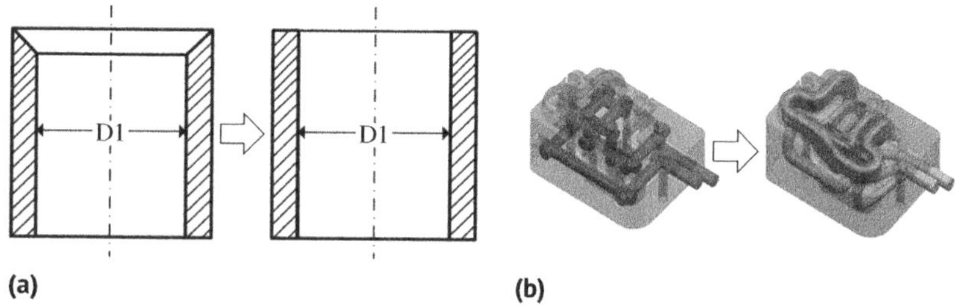

Figure 4.18 Two scenarios of FI redesign. (a) Simplification of an existing FI and (b) synthesis of a new FI.

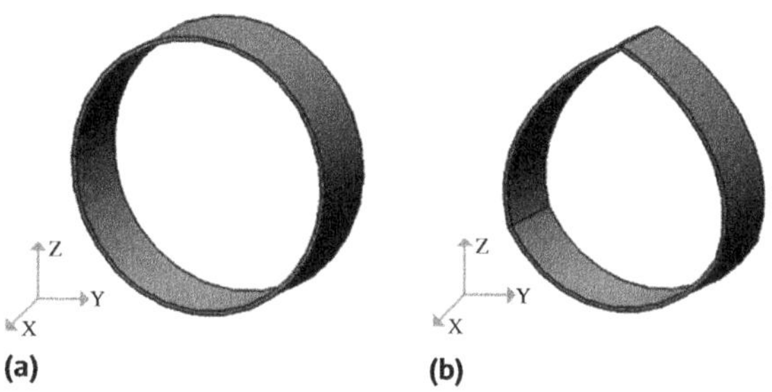

Figure 4.19 An illustrative example of FIs needing support structure or being self-supported. (a) Circular cross section and (b) teardrop cross section.

4.5.2.3 Determination of Building Orientation

BO refers to the material adding direction and it is normally perpendicular to the building plate. BO is the determining factor of whether an FI needs support structure. For some special FIs especially internal channels, removal of support structure is still a challenge. As such, self-support FIs are highly recommended although the optimal functional performance may be compromised. One example of depicting the effect of BO on the shape of FI is presented in Figure 4.19. The BO is along Z axis so that the circular FI must be modified as teardrop shape to avoid support structure.

As summarized in [50], intensive studies have been reported to optimize BO with various objectives including the minimization of surface roughness, build time, need of supports, distortion, and the increase of part stability. Different from the general practice of finding the optimal BO for the whole part, design for performance requires the successful build of FIs, especially critical FIs. Therefore, the optimal BO needs to fully consider the following rules in sequence.

1. Part-environment FIs are of the highest priority to be support-free because these FIs deliver the primary functions.
2. Part-part FIs are of the second priority because they are subject to no change; otherwise, the mating interface with external parts needs to be redesigned, which is normally unacceptable.
3. Part-user FIs are of the third priority because they are more flexible.
4. If the accessibility of support structure of any of the above three is feasible, the corresponding rule can be neglected accordingly.
5. If there are conflicts because multiple FIs in the same type prefer different BOs, the optimal BO is determined by the largest number of FIs in the same BO. If it is still ineffective, a

compromised BO should be calculated fully considering the whole set of FIs in the same type. For example, the centroid line of FIs in a circular pattern.

4.5.2.4 Construction of FV and IDS

The constructions of FV and IDS are correlated. As introduced in Section 4.5.1, a FV is constructed by thickening the FI with a proper thickness. This thickness depends on the specific method of how to generate IDS. There are two methods for generating the IDS: connect-the-dots (CTD) method and maximum boundary (MB) method. CTD method is inspired by Ullman's statement [51] that fewer than 20% of the dimensions on most components are critical to performance and designing the body is often a sophisticated CTD problem once FIs are defined. The MB method is to use the allowable maximum dimensions to define a body to encapsulate all predefined FVs. The details of these methods and applicable scope are elaborated as follows by using two separate FVs shown in Figure 4.20a.

As for the CTD method, the basic idea is to define the FVs as precise as possible and then heuristically uses linking material to connect discrete FVs as an integrated body. The way of defining the links is highly dependent on the designers' experience and FEA results. However, the fundamental principle of the CTD method is to construct a lightweight IDS with barely minimum material. As shown in Figure 4.20b and c, discrete FVs are modified with appropriate dimensional parameters and then joined by minimum LV to form the IDS. The generated IDS is then validated against loading conditions and passed to downstream refinement process only for improving manufacturability.

For the MB method, it is inspired by the division of design space and non-design space in topology optimization study. Consistent with this division, FVs should be treated as non-design space while the subtraction product of IDS and FVs should be regarded as design space. This method brings two benefits. First, thickness specification of FVs becomes more flexible. This is because the IDS is recommended to be as large as maximum boundaries allow; therefore, even if the initial thickness of an FV is smaller than the real one as requested by local stress, extra material will be added to the FV to satisfy local stress. Second, applying topology optimization for IDS can provide multiple

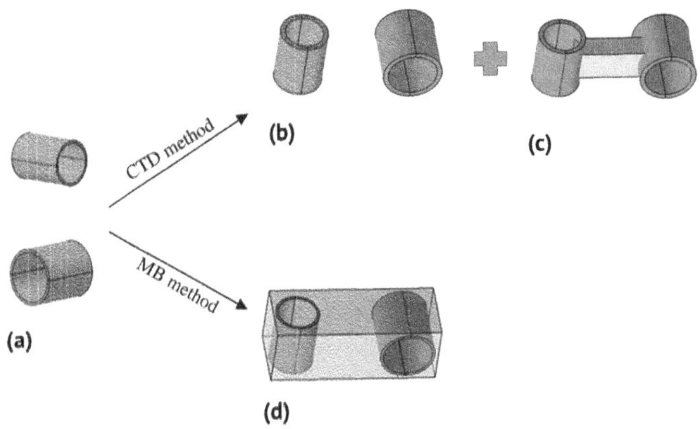

Figure 4.20 An illustrative example of the connect-the-dots method and maximum boundary method. (a) Discrete FVs, (b) refine the thickness of FVs, (c) define the IDS, and (d) define the IDS.

feasible solutions for design inspiration and it is possible to find the optimal linking material distribution for the lightweight purpose. As shown in Figure 4.20d, the FVs are captured by a volume defined by the maximum boundary.

In general, the preference over CTD or MB method depends on two criteria: number of FIs and loading criticality of LV. The applicability of CTD and MB is compared in Table 4.4. When there exists a large number of FIs, it is recommended to apply CTD method regardless of the loading criticality of the LV. This is because the CTD method does not need to follow the long design cycle of MB, which includes IDS generation, topology optimization, redesign, and FEA validation; therefore, it reduces the product lead time. Besides, excessive FIs always mean that the loading conditions are highly complex so that it is challenging for topology optimization to find the optimal solution. As shown in the example of hydraulic block manifold [52] in Figure 4.21a, CTD method is more suitable to directly bind these the excessive number of channels because the LV is not load-critical. In contrast, when there are only a few FIs, the MB method offers extraordinary capabilities to synthesize innovative LV. Moreover, it offers the possibility of design automation from the steps of FI definition, IDS generation, and TO. One control arm example [53] is shown in Figure 4.21b, where a more creative design solution can be found via the MB method.

Table 4.4 The preference table of CTD and MB methods.

		Number of FIs	
		Small	**Large**
Loading criticality	Low	Both	CTD
	High	MB	CTD

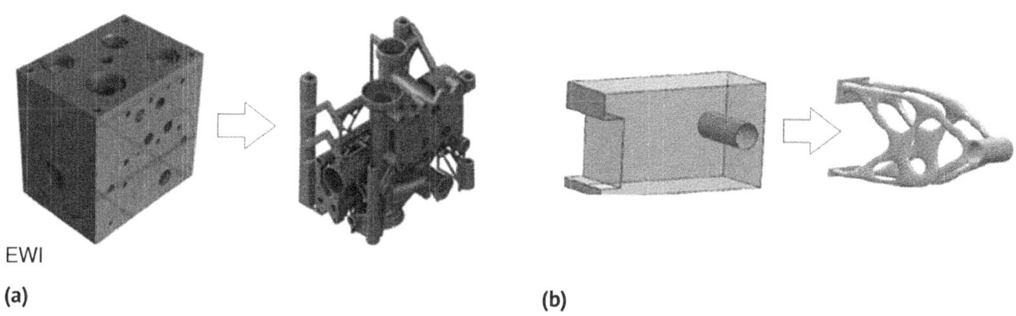

Figure 4.21 Two illustrative examples of IDS generation via applying CTD and MB methods. (a) Hydraulic block manifold redesign with CTD method and (b) control arm with MB method.

4.5.3 Revisiting Throttle Pedal Example

The throttle pedal example is revisited here to demonstrate how the functional entity-based design method works for fusion of parts. There were two groups of part candidates identified in Section 4.4.4, and the part group (C2, C3, C4, C5, C8) is chosen as an exploratory example. The details are elaborated in the following sections.

4.5.3.1 Functional Analysis

The primary goal of this step is to establish function-FI mappings on the basis of function-part mappings that have already been built in Figure 4.10. Taking the hollow shaft (C5) as an example (see Figure 4.22a), the detailed steps are as follows.

1. Part-environment FI. Since the shaft does not need to interact with any external environment that does not belong to the throttle pedal system, there exists no such FI in the part C5.
2. Part-user FI. Similar as Step 1, this part does not have direct contact with users; therefore, there exists no such FI as well in part C5.
3. Part-part FI. This part has external interactions with the bearing (C6), retaining ring (C7), right casing (C4), rotation limit (C10), and torsional spring (C11). As depicted in Figure 4.22b, the primary functions delivered by this part include "allow rotation" and "constrain the rotation." These two functions are corresponding to the FIs that couple the bearing and rotation limit, respectively. These FIs are marked as No. 6 and No. 4 in Figure 4.22b. The realization of the function "allow rotation" needs a supportive function "reduce the translation DoF" via the retaining ring. As such, the FI is extracted and marked as No. 5. Moreover, the shaft also needs to position, guide, and fix the torsional spring; thus, three more FIs are extracted as shown in Figure 4.22b as No. 3, 2, and 1, respectively.
4. System boundaries (SBs). The left end of the part C5 needs to be well bounded to secure the successful insertion into the space defined by the left casing. Therefore, feature No. 4 is extracted to define the maximum dimensions of part C5 as shown in Figure 4.22b.
5. Manufacturing or assembly features. This step is necessary to examine the design intention of each feature to avoid ignorance of any FI that delivers a function. Other features such as undercut, tapered holes for assembly, and lightweight cut-out are normally not FIs.

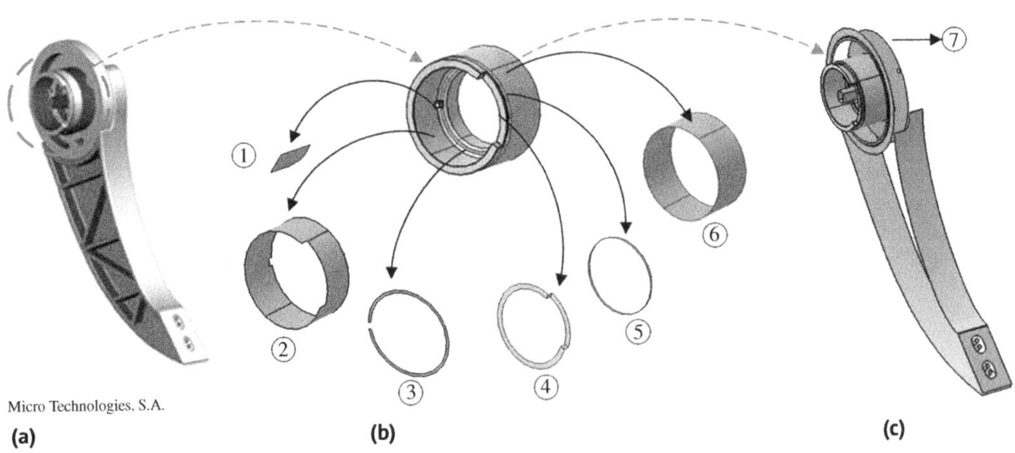

Figure 4.22 Extract FIs and SBs and establish function-FI mappings. (a) Candidacy assembly design, (b) hollow shaft FI decomposition, and (c) all FIs and SBs. (Please refer to the online version for the color image.)

Repeat the above steps and all FIs can be identified as well as the corresponding functions. The FIs and SBs are summarized in Figure 4.22c. Part-environment FIs, part-part FIs, and SBs are colored in green, blue, and yellow, respectively. In this example, there is only one part-environment FI as shown in Figure 4.22c. Feature No. 7 is to prevent dust. These extracted FIs are subject to modification before being used for constructing FVs.

4.5.3.2 Synthesis of New FIs

Functional interfaces in this case study are primarily defined for coupling purpose, which is highly bounded by the mating parts and leaves no room for performance improvement or synthesizing a new FI. Therefore, the main emphasis in this section is to generalize the shape or alter the parameters of extracted FIs. There are mainly three scenarios. First, extracted FIs may have obsolete manufacturing or assembly features. For example, the primary function of the flat surface is to prevent dust in Figure 4.23a. The holes created on the surface are assembly holes for screws to connect the shaft and lever. Therefore, the holes are obsolete and supposed to be removed. A generalized FI is presented in Figure 4.23b.

Second, resizing of extracted FIs is one of the common requirements for new FIs because consolidation may contribute to a more compact design. As shown in Figure 4.23c and d, the length of the cylindrical surface is shortened from L1 to L2 to define the real section that accommodates the bearing. Third, there is a need to rejoin discrete FIs. For example, deletion of round corners in the grove that was designed for milling purpose results in disconnected surfaces (gaps can be found between surfaces, as shown in the circled area A in Figure 4.23c). The real dimension of the desired FI is L3 as colored as yellow in Figure 4.23d. Each FI should be examined whether the dimensions, topology, and completeness of comprised surfaces are right as demanded.

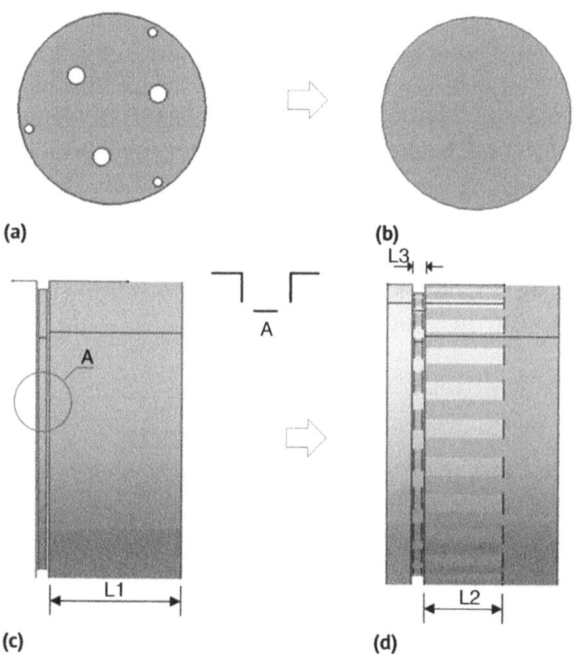

Figure 4.23 Modification of extracted FIs. Note: The strips are originally colored in yellow representing the real dimensions. (Please refer to the online version for the color image.)

4.5.3.3 Determination of Building Orientation

The layout and shape of FIs determine the optimal BO, which secures the successful build of FIs, especially those with high surface quality requirements. The main strategy is to analyze the priority of each FI. The priority is determined by three factors: 1) whether the FI delivers the primary function of the part so that as-fabricated geometry should be exactly matched with as-design one, 2) whether the FI needs support structure to prevent distortion, and 3) whether post-processing of the FI is viable. The detailed procedure is presented as follows.

1. BO for the part-environment FI. As shown in Figure 4.24, Feature 1 is to prevent dust and the flat surface allows easy post-processing even if support structure is added.
2. Building for the part-part FI. Most part-part FIs are comprised of cylindrical surfaces except the flat FI marked as Feature 4 in Figure 4.24. To guarantee the cylindricity of these FIs, the BO must be parallel to the centroid line of these FIs. For example, Feature 2 must be cylindrical to mate with the bearing and any other BO may lead to excessive rework. Special attention is drawn for Feature 3 because the top surface of the groove is overhanging.
3. Other nonfunctional criteria. Through the above two steps, there still exist two optional BOs: along positive Z axis or negative Z axis. Then, other economic considerations such as material consumption and building time should be applied. Since this part is simple enough, it can be easily observed that the optimal BO should be the positive Z axis, as indicated by the arrow in Figure 4.24.

Re-examination of each FI with the selected BO also indicates that there is no need to alter the shape of FI to be support-free.

4.5.3.4 Construction of FV and IDS

An FV can be constructed by thickening the FS along its normal direction. The thickness of each FI may be different and it must consider the function to be delivered and local stress. As shown in Figure 4.25a, the FI is formed by two sub-FIs, which accomplish functions of "couple retaining ring" and "couple bearing," respectively. Ideally, the thickness of these two FIs should be different because the local stress is much higher for the latter one. However, current geometry modelling tools fall short in offering the flexibility of assigning various thickness values for different sections; therefore, both FIs are assigned the same thickness at 0.5 mm as shown in Figure 4.25b and c, respectively. Following the same strategy, all FSs are assigned essential thickness to form FVs. It should be advised that the selection of thickness value should be relatively small, which grants designers more freedom to explore the linking patterns amongst FVs.

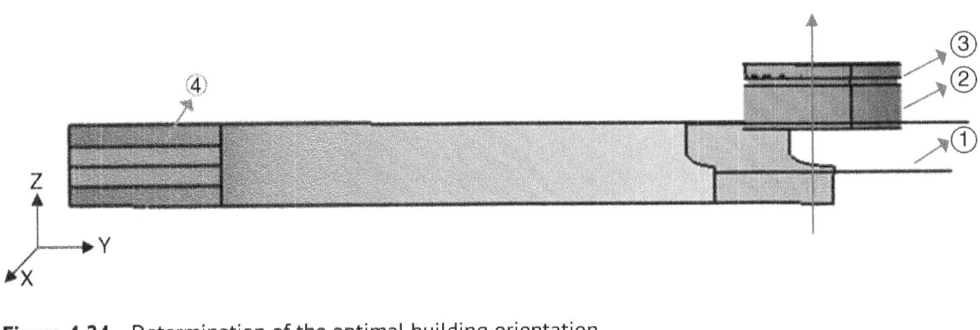

Figure 4.24 Determination of the optimal building orientation.

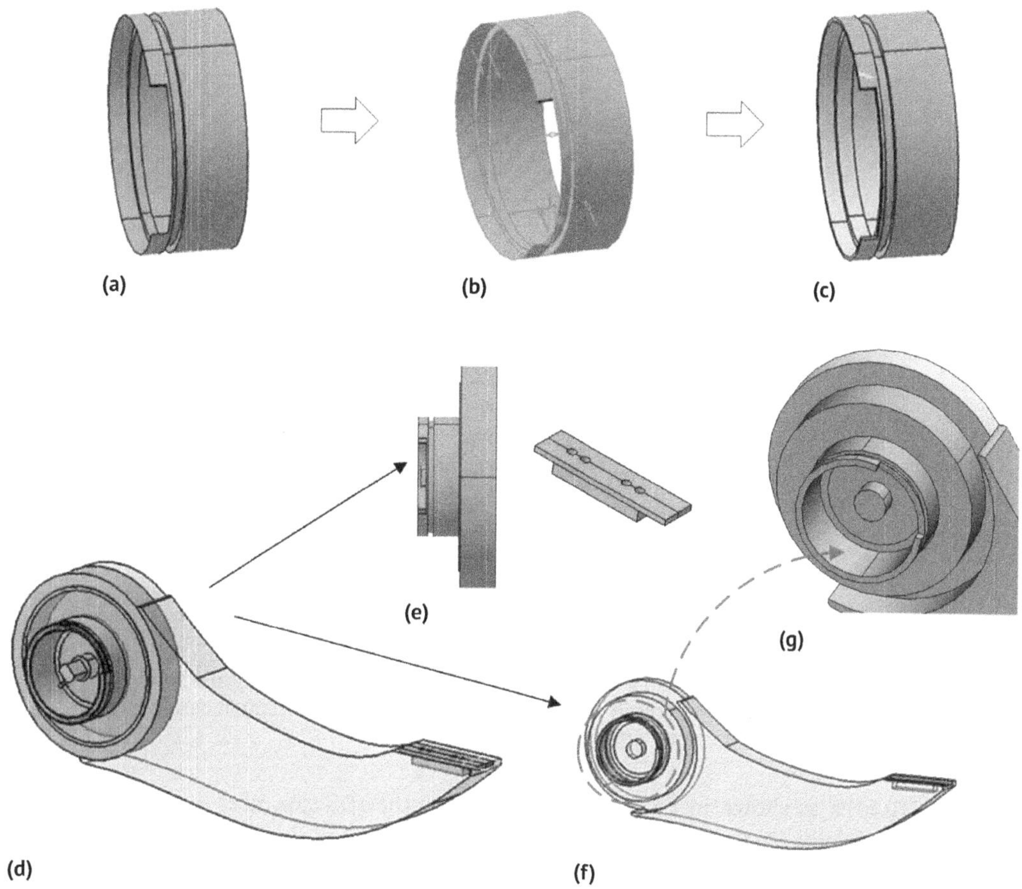

Figure 4.25 Construction of FV and IDS. (a) Functional interface, (b) thickening direction, (c) functional volume, (d) initial design space, (e) non-design space, (f) design space, and (g) detailed view.

In this example, there exist only a small number of FVs and the LV is load-critical; therefore, the MB method is applied instead of the CTD method. All FVs are encapsulated in an IDS which is conformal to all FIs and confined by the SBs. The generated IDS is displayed in Figure 4.25d. FVs and LV are colored differently and the LV can be easily computed by subtracting FVs from the IDS. In accordance with topology optimization, the FVs and LV can be defined as non-design space and design space, respectively. In this way, the design freedom of topology complexity can be employed to synthesize the optimal material distribution within the IDS. Here, we can see the IDS is a functionally equivalent solution to the original assembly design.

4.5.3.5 Refinement of the Constructed IDS

To test the constructed IDS against loading conditions, FEA is conducted to make sure the IDS can satisfy mechanical requirements as well. As shown in Figure 4.26a, the IDS is fixed by a single-point constraint inside the hollow shaft area and a pressure load of 1120 N is applied on the pedal. FEA result indicates that the maximum stress is 40 MPa lower than the yield

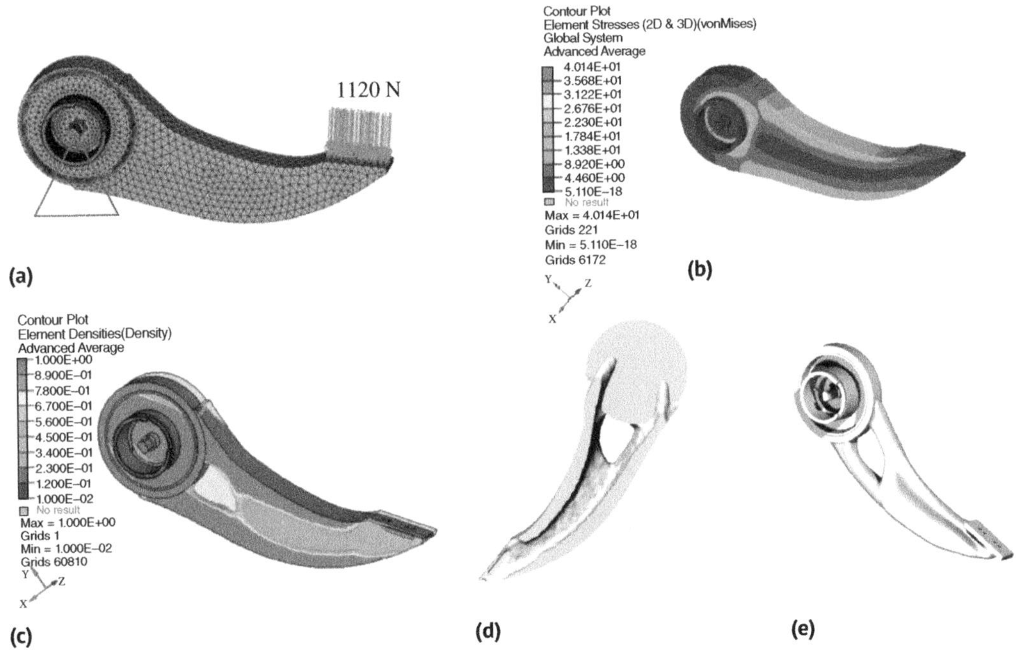

Figure 4.26 Refinement of the IDS and a final consolidated design solution. (a) FEA model, (b) stress distribution, (c) topology optimization result with 30% volume fraction, (d) conceptual design (e) final design solution with manufacturability considerations.

stress of 210 MPa as shown in Figure 4.26b. Therefore, the IDS can meet the functional and loading requirements. To further investigate the potential of lightweight, topology optimization is conducted in OptiStruct [54] with FVs and the subtraction product of IDS and FVs as non-design space and design space, respectively. With the volume fraction constraint of 30% and the objective of minimizing compliance, the optimized result is depicted in Figure 4.26c. The portion represented in red indicates the relative density of 1 and the blue represents void. Then, the topologically optimized result is exported and smoothed. The back view of the smoothed conceptual design is shown in Figure 4.26d. With further consideration of manufacturing constraints and anticorrosion requirements, the linking surfaces are reconstructed to avoid the tiny organic features. One feasible final solution is shown in Figure 4.26e with 20.3% weight reduction. This case study proves the value of the functional entity in aiding the reconstruction process of consolidated design as well as supporting direct applicability in downstream activities such as topology optimization.

4.6 Emerging Trends in Part Consolidation and Assemblies

Apart from tackling the challenges of part candidate filtering and fusion of parts, research on assembly design with AM has also witnessed some interesting progress. Representative efforts include the automation of part filtering methods and assembly-free design.

4.6.1 Automated Part Filtering Methods for Part Consolidation

The efficiency of manually applying the candidacy rules and principles for part filtering becomes a concern when the system of interest is very complex. The need of developing an automated tool has been discussed by several scholars. These ongoing works can be classified into two groups: function-driven approach [55] and cost-driven approach [56]. The former approach targets at clustering part candidate groups according to the functional proximity between parts and candidacy rules, while the latter one emphasizes on building cost under different numbers of clusters.

Yang et al. [55] automated the part filtering process of with the 7 candidacy rules and 5 principles covered in this chapter for a drone with 232 parts. In their work, functional and physical interaction graph is first constructed. Then modularization and community detection are optimized by maximizing the modularity, and verification of candidacy rules is automatically implemented in MATLAB. As shown in Figure 4.27, thirty part candidate groups are identified out of the drone model. Nie et al. [56] use minimal cost as the objective to find the optimal number of part groups for consolidation. In their approach, similar steps are taken to build the adjacency graph of the assembly and find the part candidates that satisfy the candidacy rules. However, they took a different way of clustering part candidates by minimizing the printing cost based on printing time, support structure and nesting layout. As shown in Figure 4.28, given different number of part clusters, the final optimal nesting layout and BO are different.

It should be advised that both research solutions showed promise in automating the part filtering process for consolidation, but they are a proof of concept and there is still a long way before any practical tools become available. The main challenges can be summarized as follows. First, obtaining the adjacency graph requires a full understanding of the functions and design requirements of the original assembly, which is a time-consuming process and needs sound background knowledge. Second, implementation of either the function-based or cost-based filtering method in existing CAD systems requires extra information (e.g., maintenance information) that is not available in current commercial CAD systems. Third, when the system of interest is overly complex, manual process of dissecting the systems at the high level is still recommended.

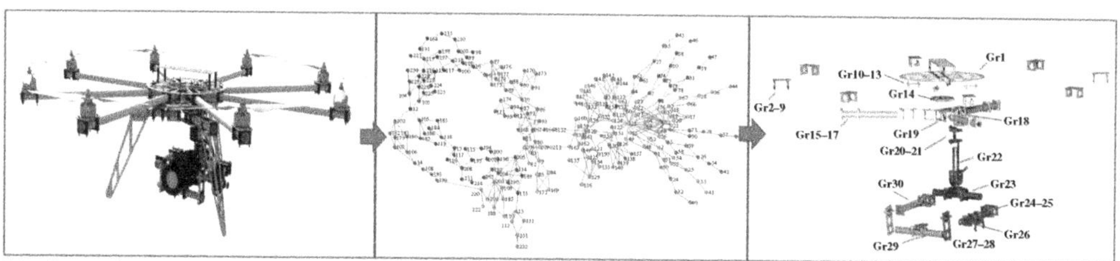

Taica Corporation

(a) (b) (c)

Figure 4.27 Automated part filtering for part consolidation. (a) Drone model with 232 parts, (b) functional and physical interaction graph of the drone, and (c) part candidate groups for consolidation for colors.

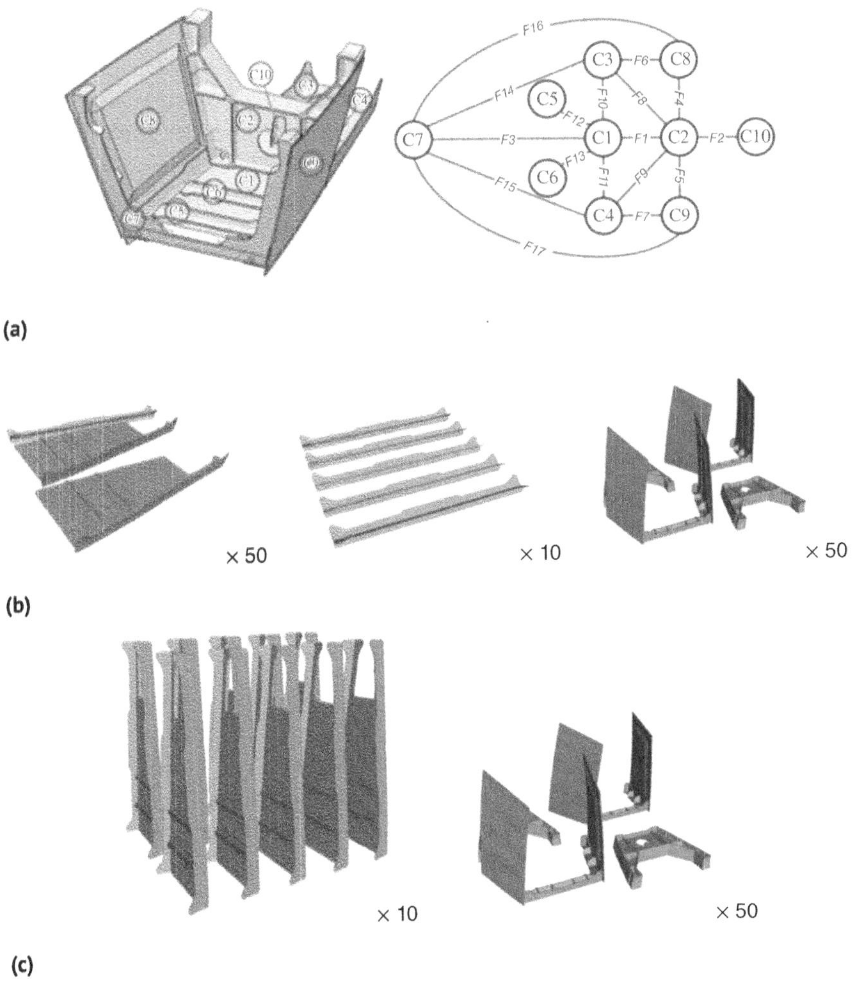

Figure 4.28 Cost-based part filtering approach with different numbers of part clusters. (a) Original assembly, (b) optimal nesting and building orientation when clusters = 4, and (c) optimal nesting and building orientation when clusters = 3.

4.6.2 Assembly-free Design and Part Segregation

Assembly-free design (also noted as non-assembly mechanism) is also an important stream of assembly design in AM community. AM allows construction of complex geometries in one single step, which permits the direct fabrication of mechanical joints that typically have relative motion between its links. Two types of assembly-free mechanisms have been reported: rigid-body joint and compliant mechanism (see Figure 4.29a and d). For rigid-body joints, they are typically comprised of multiple rigid bodies, and the joint performance is highly dependent on the surface quality and clearance in between. As such, the challenges for rigid-body joints include the use of support structure and removal of trapped materials in the building process [7]. To accommodate these requirements, traditional joints must be redesigned with new shapes/topologies that help to improve the surface quality. For example, traditional cylindrical pin must be redesigned to be

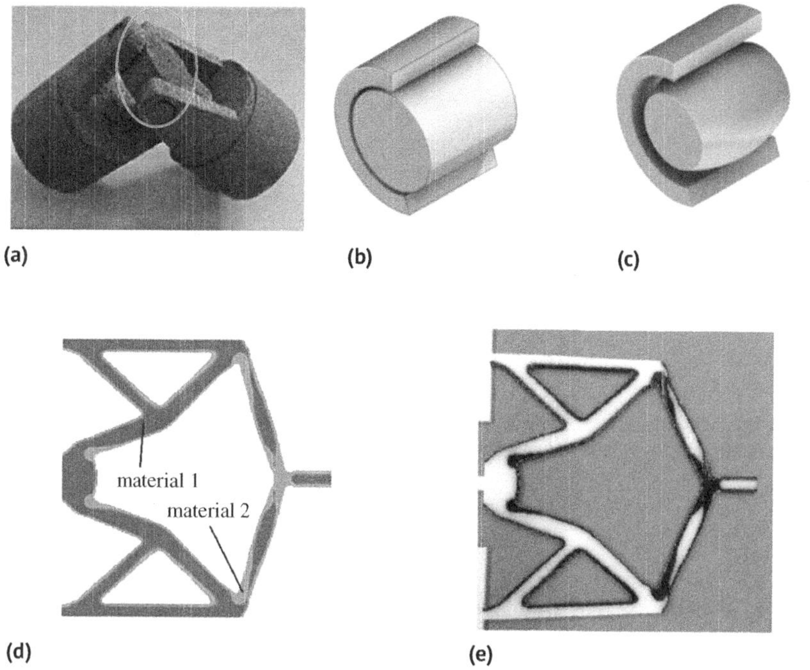

Figure 4.29 Assembly-free mechanisms fabricated by AM. (a) Universal joints, (b) traditional pins, (c) drum-shaped pin, (d) displacement inverter (as design) and (e) displacement inverter (as fabricated).

drum-shaped pin to increase the clearance between the sleeve and the pin so that trapped powder can be easily removed as shown in Figure 4.29b and c. Compliant mechanisms allow relative motion of elements through deformation of elastic members; therefore, there is no clearance requirement for its links. Instead, the material properties and geometric configuration of the mechanism play a critical role in determining the performance. As shown in Figure 4.29d and e, a displacement inverter can be achieved by a compliant mechanism made of two materials [57]. Although both rigid and compliant joints have been proven feasible for using AM, these applications suffer from inaccurate precision motion and do not allow large deformation and long-term service life. Potential applicable fields of the assembly-free mechanisms can be micro-scale valves/pumps that do not require precision motion control and where traditional mechanisms are bounded by its physical sizes.

Part segregation (also noted as part decomposition) aims at dissecting a part into assemblies for ease of manufacturing, reduction of printing time, or overcoming size limitations of 3D printers. As opposed to PC, the number of part count increases after part segregation. Some examples are shown in Figure 4.30. Typical challenges to be solved in part segregation include 1) part decomposition (how many pieces and where to cut), 2) determination of the best BO, 3) determination of the nesting layout to minimize printing time and material cost, and 4) assembly interface design (e.g., interlocking mechanisms). Current applications of part segregation are limited to non-loading critical examples such as toys and prototypes to reduce printing time. As pointed out in literature [58], further studies are needed to automate the part decomposition process [59] and incorporate design for assembly techniques.

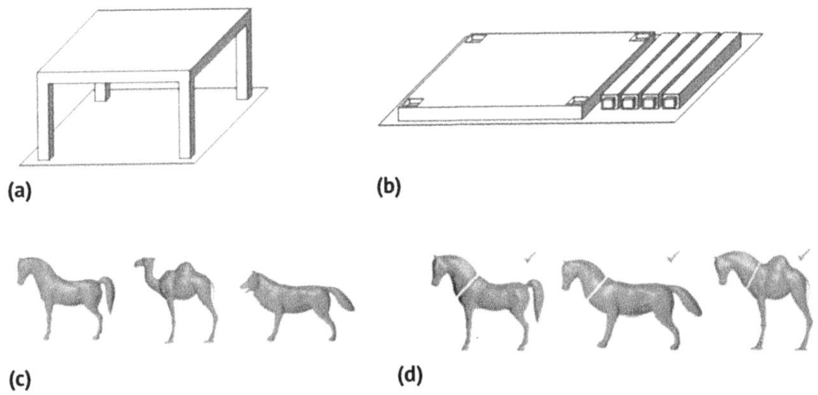

Figure 4.30 Examples for the need of part segregation. (a) A table, (b) optimal printing setup, (c) a list of animals, and (d) interchangeable parts customized for different animal models.

4.7 Summary

This chapter introduces a holistic PC design method to support the identification of part candidates for consolidation and the fusion of parts. Seven candidacy rules and five principles are included to help selectively find the part groups, which are suitable for PC. A functional entity-based method is also elaborated to aid the process of converting part candidates into an integrated one-piece design. A case study of a throttle pedal example is conducted with detailed steps to wire the core ideas in practical applications. Since AM processing capability keeps evolving, more and more process constraints can be overcome such as quality, time, and material availability so that the candidacy rules should be updated and consistent with the AM facilities that are at hand. In the near future, PC techniques are expected to be applied to more sophisticated engineering systems with automated approaches. Meanwhile, the progress in assembly-free design and part segregation will contribute to more parts that can be printed with AM process.

References

1 Boothroyd, G., *Product design for manufacture and assembly*. 2002, CRC Press: New York, USA.
2 Oxford English dictionary. *Definition of consolidation*. 2018; https://en.oxforddiction-aries.com/definition-/consolidation.
3 Anderson, D.M., *Design for manufacturability: how to use concurrent engineering to rapidly develop low-cost, high-quality products for lean production*. 2014, CRC Press.
4 Dewhurst, N.P., DFMA the product, then lean the process. in *Proceedings of the International Forum on Design for Manufacture and Assembly*, 2010. pp. 15–16.
5 Lipson, H., et al., *3-D printing the history of mechanisms*. Journal of Mechanical Design, 2005. **127**(5): pp. 1029–1033.
6 Calì, J., et al., *3D-printing of non-assembly, articulated models*. ACM Transactions on Graphics (TOG), 2012. **31**(6): p. 130.
7 Cuellar, J.S., et al., *Additive manufacturing of non-assembly mechanisms*. Additive Manufacturing, 2018. **21**: pp. 150–158.

8 Kataria, A. and Rosen, D.W., *Building around inserts: methods for fabricating complex devices in stereolithography*. Rapid Prototyping Journal, 2001. **7**(5): pp. 253–261.
9 Panesar, A., et al., *Design framework for multifunctional additive manufacturing: placement and routing of three-dimensional printed circuit volumes*. Journal of Mechanical Design, 2015. **137**(11): 111414.
10 GE Capital. *Additive manufacturing redefining what's possible*. 2013; Available from: http://www.americas.gecapital.com/-GECA_Document/Additive_Manufacturing_Fall_2013.pdf.
11 Dietrich, D.M. and Cudney, E., *Impact of integrative design on additive manufacturing quality*. International Journal of Rapid Manufacturing, 2011. **2**(3): pp. 121–131.
12 Ford, S. and Despeisse, M., *Additive manufacturing and sustainability: an exploratory study of the advantages and challenges*. Journal of Cleaner Production, 2016. **137**: pp. 1573–1587.
13 Lu, Y., et al., Self-improving additive manufacturing knowledge management. in *ASME 2018 International Design Engineering Technical Conferences and Computers and Information in Engineering Conference*, 2018. p. V01BT02A016.
14 Suh, N.P., *The principles of design*. 1990, Oxford University Press: New York, (USA).
15 Savransky, S.D., *Engineering of creativity: Introduction to TRIZ methodology of inventive problem solving*. 2000, CRC Press.
16 Fagade, A.A. and Kazmer, D., Optimal component consolidation in molded product design, in *International Design Engineering Technical Conferences and Computers and Information in Engineering Conference* . 1999. Vol. 19746, pp. 255–265. Citeseer: Las Vegas, Nevada.
17 Frey, D., et al., *Part count and design of robust systems*. Systems Engineering, 2007. **10**(3): pp. 203–221.
18 Lowe, G. and Bogue, R., *Design for disassembly: a critical twenty-first century discipline*. Assembly Automation, 2007. **27**(4): pp. 285–289.
19 Schmelzle, J., et al., *(Re) Designing for part consolidation: understanding the challenges of metal additive manufacturing*. Journal of Mechanical Design, 2015. **137**(11): p. 111711.
20 Gibson, I., Rosen, D.W., and Stucker, B., *Additive manufacturing technologies: rapid prototyping to direct digital manufacturing*. 2010, Springer: US.
21 Klahn, C., Leutenecker, B., and Meboldt, M., *Design strategies for the process of additive manufacturing*. Procedia CIRP, 2015. **36**: pp. 230–235.
22 Yang, S., Tang, Y., and Zhao, Y.F., *A new part consolidation method to embrace the design freedom of additive manufacturing*. Journal of Manufacturing Processes, 2015. **20**: pp. 444–449.
23 Report, G.E., *An epiphany of disruption: GE additive chief explains how 3D printing will upend manufacturing*. 2017; http://www.ge.com/reports/epiphany-disruption-ge-additive-chief-explains-3d-printing-will-upend-manufacturing/.
24 Ponche, R., et al., *A new global approach to design for additive manufacturing*. Virtual and Physical Prototyping, 2012. **7**(2): pp. 93–105.
25 Leutenecker-Twelsiek, B., Klahn, C., and Meboldt, M., *Considering part orientation in design for additive manufacturing*. Procedia CIRP, 2016. **50**: pp. 408–413.
26 Orme, M.E., et al., *Designing for additive manufacturing: lightweighting through topology optimization enables lunar spacecraft*. Journal of Mechanical Design, 2017. **139**(10): pp. 100905–100906.
27 Thompson, M.K., et al., *Design for additive manufacturing: trends, opportunities, considerations, and constraints*. CIRP Annals-Manufacturing Technology, 2016. **65**(2): p. 737–760.
28 Booth, J.W., *Comparing functional analysis methods for product dissection tasks*. Journal of Mechanical Design, 2015. **137**(8): p. 081101.

29 Nagel, R.L., Bohm, M.R., and Linsey, J.S., An investigation into the effectiveness of an algorithmic approach to teaching functional modeling. in *International Design Engineering Technical Conferences and Computers and Information in Engineering Conference*, 2013. American Society of Mechanical Engineers: Portland, USA. p. V001T004A023.

30 Yang, S., Santoro, F., and Zhao, Y.F., *Towards a numerical approach of finding candidates for additive manufacturing-enabled part consolidation*. Journal of Mechanical Design, 2018. **140**(4): p. 041701.

31 Yang, S., Rules and methods for exploring and utilizing additive manufacturing-enabled part consolidation potentials in product redesign. PhD thesis. 2019. McGill University (Canada).

32 Ulrich, K., *The role of product architecture in the manufacturing firm*. Research Policy, 1995. **24**(3): pp. 419–440.

33 Hölttä-Otto, K. and De Weck, O., *Degree of modularity in engineering systems and products with technical and business constraints*. Concurrent Engineering, 2007. **15**(2): pp. 113–126.

34 Stone, R.B., Wood, K.L., and Crawford, R.H., *A heuristic method for identifying modules for product architectures*. Design Studies, 2000. **21**(1): pp. 5–31.

35 Gershenson, J., Prasad, G., and Zhang, Y., *Product modularity: definitions and benefits*. Journal of Engineering Design, 2003. **14**(3): pp. 295–313.

36 Hölttä-Otto, K., et al. *Comparative analysis of coupling modularity metrics*. Journal of Engineering Design, 2012. **23**(10-11): pp. 790–806.

37 Lipson, H., *Principles of modularity, regularity, and hierarchy for scalable systems*. Journal of Biological Physics and Chemistry, 2007. **7**(4): p. 125.

38 Yu, T.L., Yassine, A.A., and Goldberg, D.E., *An information theoretic method for developing modular architectures using genetic algorithms*. Research in Engineering Design, 2007. **18**(2): pp. 91–109.

39 Jung, S. and Simpson, T.W., *New modularity indices for modularity assessment and clustering of product architecture*. Journal of Engineering Design, 2017. **28**(1): pp. 1–22.

40 Sosa, M.E., Eppinger, S.D., and Rowles, C.M., *A network approach to define modularity of components in complex products*. Journal of Mechanical Design, 2007. **129**(11): pp. 1118–1129.

41 Boer, M. and Logendran, R., *A methodology for quantifying the effects of product development on cost and time*. IIE Transactions, 1999. **31**(4): pp. 365–378.

42 Lucchetta, G., Bariani, P., and Knight, W., *Integrated design analysis for product simplification*. CIRP Annals-Manufacturing Technology, 2005. **54**(1): pp. 147–150.

43 Bell, J., Snooke, N., and Price, C., *A language for functional interpretation of model based simulation*. Advanced Engineering Informatics, 2007. **21**(4): pp. 398–409.

44 Kitamura, Y., Koji, Y., and Mizoguchi, R., *An ontological model of device function: industrial deployment and lessons learned*. Applied Ontology, 2006. **1**(3): pp. 237–262.

45 Tang, Y., A multiscale and multifunctional design method for mesoscale lattice structures fabricated by additive manufacturing processes. PhD thesis. 2017. McGill University.

46 Hirtz, J., et al., *A functional basis for engineering design: reconciling and evolving previous efforts*. Research in Engineering Design, 2002. **13**(2): pp. 65–82.

47 Kusiak, A., *Computational intelligence in design and manufacturing*. 2000, John Wiley & Sons.

48 Chakrabarti, A., et al., *Computer-based design synthesis research: an overview*. Journal of Computing and Information Science in Engineering, 2011. **11**(2): p. 021003.

49 Autodesk. *Project dreamcatcher*. 2016; https://autodeskresearch.com/projects/dreamcatcher.

50 Moroni, G., Syam, W.P., and Petrò, S., Functionality-based part orientation for additive manufacturing. Procedia Cirp, 2015. **36**: pp. 217–222.

51 Ullman, D.G., *The mechanical design process*. 1992, McGraw-Hill New York.

52 Renishaw, *Hydraulic block manifold redesign for additive manufacturing*. 2016; https://www.-qualitymag.com/articles/94507-hydraulic-block-manifold-redesign-for-additive-manufacturing.

53 Ponche, R., Méthodologie de conception pour la fabrication additive, application à la projection de poudres. 2013, Ecole centrale de nantes-ECN.

54 Altair, *OptiStruct Overview*. 2018; https://altairhyperworks.com/product/OptiStruct.

55 Yang, S., et al., *A numerical-based part consolidation candidate detection approach with modularization considerations*. Research in Engineering Design, 2019. **30**(1): pp. 63–83.

56 Nie, Z., et al., *Optimization of part consolidation for minimum production costs and time using additive manufacturing*. Journal of Mechanical Design, 2020. **142**(7): 1–16.

57 Sigmund, O., *Design of multiphysics actuators using topology optimization–Part I: one-material structures*. Computer Methods in Applied Mechanics and Engineering, 2001. **190**(49-50): pp. 6577–6604.

58 Oh, Y., Zhou, C., and Behdad, S., *Part decomposition and assembly-based (Re) design for additive manufacturing: a review*. Additive Manufacturing, 2018. 22: pp. 230–242.

59 Maiyar, L.M., et al., *Part segregation based on particle swarm optimisation for assembly design in additive manufacturing*. International Journal of Computer Integrated Manufacturing, 2019. **32**(7): pp. 705–722.

5

Manufacturing Constraints and Design Guides for Additive Manufacturing

In the dynamic landscape of additive manufacturing (AM), the expansion of design freedom for complex geometries has been a game-changer. However, this innovation is not without its challenges. AM technologies, despite their advancements, encounter specific manufacturing limitations that can significantly impact factors such as fabrication cost, time, quality, and even the feasibility of printing certain designs. Recognizing and adapting to these limitations during the initial design stages is not just beneficial but necessary.

This chapter delves into the intricate world of design rules and guidelines tailored for AM, paralleling the traditional methodologies of design for manufacturing and assembly. These guidelines are not mere suggestions; they are critical directives that shape the geometry and material choices, ensuring designs are not only practical but also economically and time-efficient for printing. This approach, sometimes referred to as restrictive design for additive manufacturing (DfAM), serves as a compass for navigating the complexities of AM design.

Our exploration here addresses the widely recognized design rules and guidelines pertinent to various AM technologies. It's crucial to understand that, unlike conventional manufacturing processes, different AM technologies—and even different machines within the same technology—have unique processing capabilities and constraints. Therefore, the design rules we discuss offer a general conceptual framework, adaptable to a range of AM technologies, while acknowledging that specific constraints like maximum overhang angles or minimum feature sizes are also influenced by factors like materials and process parameters. The values and suggestions provided here should be seen as a general reference, with the understanding that testing and benchmarking on selected machines and materials are essential for precise values on each constraint.

To provide a comprehensive overview, this chapter is divided into two main sections. Section 5.1 will offer a general introduction to the overarching design constraints in AM, setting the stage for deeper analysis. Following this, we will engage in a detailed examination of the design guides for specific AM processes, unraveling the rationale behind these constraints and their practical implications.

Through this chapter, we aim to equip designers and engineers with the knowledge and tools necessary to harness the full potential of AM, transforming limitations into opportunities for innovation and efficiency.

5.1 General Constraints and Guidelines for AM

Regardless of the specific AM process chosen, there are universal constraints that all of them share. This section aims to provide designers with a comprehensive understanding of these limitations. By summarizing the general constraints and guidelines that are integral to AM processes, we offer valuable insights right from the initial stages of design. These considerations do more than just improve manufacturability; they are crucial in guiding designers toward the selection of appropriate materials and processes, and in reducing fabrication time and costs.

In the subsequent parts of this section, we will methodically explore each of these constraints. Based on each constraint, some design guidelines will be introduced. Our discussion will be structured to provide a clear, in-depth understanding of how each constraint impacts the AM process and the strategies to navigate these challenges effectively.

5.1.1 General Constraint 1: Printing Volume

Printing volume stands as one of the most fundamental constraints that designers must consider in AM. Each type of AM process has inherent limitations when it comes to the size of parts they can produce. For machines that utilize a Cartesian motion system, these dimensional constraints are often straightforward and can be defined by the dimensions of the bounding box. For instance, the typical build volume for a machine employing material extrusion (ME) or fused filament fabrication (FFF) is approximately 250 mm × 210 mm × 210 mm. Such specifics are generally available in the technical specifications provided by machine manufacturers.

It's important to recognize that the factors limiting the print size of parts vary across different AM processes. Generally, machines based on extrusion or directed energy deposition (DED) are preferred for fabricating larger parts. These processes are advantageous because they require movement of only the nozzle or print head within a large workspace, facilitating the printing of sizable components. Such capabilities make these processes ideal for producing large-scale items like aircraft frames, automotive vehicle bodies, and even entire houses.

In contrast, powder bed-based processes, such as binder jetting and powder bed fusion (PBF), face more challenges in printing large parts. The primary obstacle is the need for a larger powder bed, which can lead to material wastage and the necessity for larger machines capable of handling the increased weight of the powder bed. This makes it less feasible and much more expensive to print large parts using these methods compared to ME or DED processes.

Below, Table 5.1 succinctly summarizes the typical printing dimensions for various AM processes, providing a comparative view of their part size capabilities. It should be noted that the size indicated in this table only means the typical size. The actual printing size may vary from machine to machine. Also, there are some large-size machines available for each process. For example, large machines [1] based on the ME process are developed to print the part size of more than 1 m in the x, y, and z directions. In this case, the cost of a machine is also higher, and it may also take much longer to print due to the availability of the machine.

To prevent designing parts that exceed the available printing volume of the chosen AM process, the following guidelines are recommended:

Design Guideline 5-1: Ensure your part fits the dimensions of the selected AM machine. If the part is oversized, consider dividing it into several sub-parts and assembling them post-printing.

Table 5.1 Comparison between different typical build volumes of different AM processes.

AM process name	Typical build volume	Cost of machine
FFF/FDM/material extrusion	300 × 300 × 300 mm	Low to medium
Powder bed fusion	300 × 300 × 400 mm	Medium to high
Binder jetting	300 × 200 × 150 mm	Medium to high
DED	1 × 1 × 1 m or even larger	High
Vat photopolymerization	145 × 145 × 175 mm	Low to medium
Sheet laminations	300 × 200 × 150 mm	Medium
Material jetting	500 × 400 × 200 mm	Medium to high

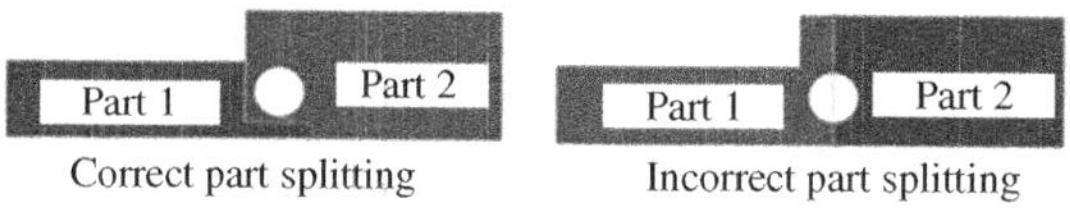

Figure 5.1 Example of part splitting.

Figure 5.1 illustrates an application of this design guideline. The part exceeds the width capacity of the chosen AM machine. Therefore, dividing the part into two sub-parts becomes necessary. During this division, care must be taken to ensure the splitting line does not intersect with critical functional features, such as the middle hole in this example, which requires precise printing for assembly purposes. Additionally, when parts are subject to specific loading conditions, it's important to design the splitting plane to avoid high-stress regions, ensuring structural integrity.

To facilitate easy assembly of the sub-parts post-printing, incorporating locating features like pins or holes during the design stage is advisable. These features aid in accurately aligning and securing the sub-parts, ensuring the assembled component functions as intended.

5.1.2 General Constraint 2: Printing Resolution, Accuracy, and Minimum Feature Size

Beyond printing volume, manufacturing resolution is a crucial consideration for designers in AM. Manufacturing resolution refers to the smallest detail variance from the design that an AM process can accurately reproduce. Typically, resolution is assessed across three axes: the Z-axis resolution is determined by layer thickness, while X- and Y-axis resolutions depend on the specific AM technology. For instance, in ME processes, XY resolution is governed by the thickness of the printing bead, which is usually around 400 μm. Its value can be adjusted through flow rate and nozzle size. Generally, designers should not expect features smaller than the resolution to be accurately produced. Along the Z axis, the resolution is mainly determined by layer thickness. For example, in the PBF process, typical layer thickness can range from 50 to 100 μm. Thus, the typical resolution for the PBF process also lies in this range, as shown in Table 5.2.

It's important to distinguish between manufacturing resolution and other concepts like printing accuracy and minimum feature size. While interrelated, these terms are distinct.

Table 5.2 Comparison of resolutions and accuracy of different AM process.

AM process name	Resolution-XY	Resolution-Z	Accuracy	Minimum feature size
FFF/FDM/material extrusion	200–400 μm	50–400 μm	±0.2 mm or ±0.5%	0.2–0.4 mm
Powder bed fusion	50–100 μm	20–100 μm	± 0.1 mm	0.4–1 mm
Binder jetting	80–200 μm	50–200 μm	±0.1–0.3 mm	0.2–0.4 mm
DED	250–500 μm	200–500 μm	±0.1–0.3 mm	1–2 mm
Vat photopolymerization	50–100 μm	25–100 μm	±0.1–0.2 mm	0.15–0.3 mm
Sheet laminations	200 μm	100–200 μm	±0.1–0.2 mm	0.2 mm
Material jetting	20–50 μm	16 μm	±0.05–0.1 mm	0.25–0.5 mm

Printing accuracy measures how closely the printed object aligns with the dimensions and specifications of the original digital model. Manufacturing resolution, however, focuses on the detail level achievable by an AM process. High printing accuracy requires not just high resolution but also control over in-process and post-process deformation. In some AM processes, such as the binder jetting process, printed parts usually suffer significant shrinkage during the post-sintering process. Thus, compensation is usually applied when the part is printed. However, it should be noted that the shrinkage rate might not be isotropic and may also cause warping or other kinds of deformation, which is not expected. All those factors will affect the accuracy of the printed part. Thus, designers need to understand the accuracy level of different AM processes, which is summarized in Table 5.2.

As to minimum feature size, it is a measure that guides designers with the minimum geometric feature the AM process can achieve. It should be larger than the machine resolution. In addition to the resolutions, other factors such as materials, and the geometric shape of features will also impact the manufacturability of features. For example, the resolution of the binder jetting process is mainly determined by its droplet size for printing nozzle. However, during printing, small features such as pins whose diameter is smaller than 0.75 mm are likely to break during the depowdering process. Thus, the minimum feature size of the binder jetting process is much larger than its resolution. Typical minimum feature sizes for different AM processes are also summarized in Table 5.2. It should be noted that those are typical values for different processes. These constraints may vary between machines, materials, and feature geometries. The geometric-specific minimum size constraints will be discussed in Sections 5.2 and 5.3.

In addition to the factors discussed above, surface roughness—defined as a measure of surface texture—also plays a crucial role in the design process. Unlike in other manufacturing methods, surface roughness in AM is influenced by a wide range of factors, including geometries, materials, and selected process parameters. Due to the complexity of these influences, a detailed discussion is beyond the scope of this section. Instead, Sections 5.2 and 5.3 will provide an in-depth examination of achievable surface roughness across different AM processes, focusing on the specific features that designers aim to produce. Based on the limitations of printing resolution, accuracy, and minimum feature size, the following design guidelines are summarized.

Design Guideline 5-2: Verify the machine's resolution, accuracy, and minimum feature size and surface roughness. If required levels are unattainable, consider post-processing and its processability.

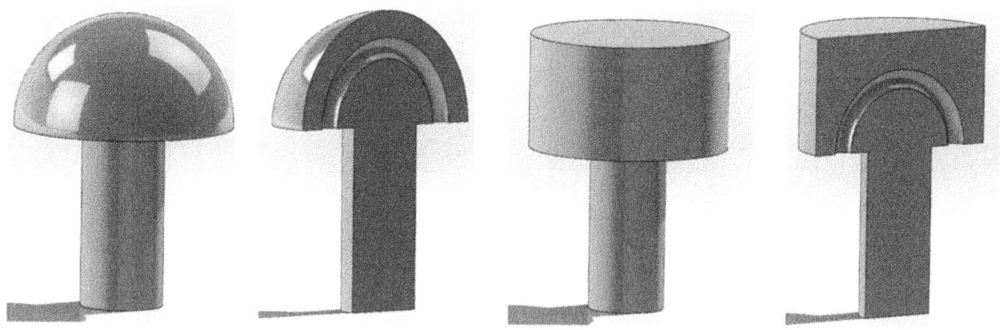

(a) (b) (c) (d)

Figure 5.2 Example of design for post-processing with the consideration of constraints of printing accuracy and surface roughness. (a) Original part, (b) section view of the original part, (c) modified part, and (d) section view of the modified part.

An example of this guideline is illustrated in Figure 5.2, which shows a mold insert with a conformal cooling channel (Figure 5.2a). This curved cooling channel design is achievable only through AM. Given that the part is made of metal, the appropriate AM method is PBF. However, to meet the functional requirements, the top spherical surface of the part must adhere to strict shape tolerance and surface roughness standards, with a maximum surface deviation of 10 μm and a surface roughness under 1.6 μm. As shown in Table 5.2, these specifications exceed the capabilities of current PBF processes. To address this, the design was modified with a flat top, allowing for post-machining to meet the high tolerance and surface finish requirements. This adjustment also enables the part to be printed without support structures.

5.1.3 General Constraint 3: Manufacturing Cost and Time

Cost and time are two additional major constraints that must be carefully considered in the design of AM. While AM can significantly reduce manufacturing preparation time, its processing time for individual units is often much longer than that of conventional manufacturing methods like injection molding or casting. Moreover, for medium- to large-scale production, traditional manufacturing methods generally remain more cost-effective. Therefore, minimizing manufacturing cost and time is a critical consideration during the design phase.

One effective approach to reducing cost and time in AM is the use of porous infills. Figure 5.3 illustrates an example of this technique, where a Stanford bunny model with a volume of 279,629 mm^3 is printed via ME with PLA at the same print speed. Compared to the solid (100%) infill in Figure 5.3b, using honeycomb or gyroid infill at a 15% rate can significantly decrease both printing time and material costs. This figure also highlights that different infill patterns and topologies yield variations in time and material consumption. Designers should therefore consider printing time, cost, and the mechanical strength and stiffness requirements when selecting an infill pattern.

The choice of infill ratio also plays an essential role. Generally, parts subjected to high or critical mechanical loads require higher infill ratios. Instead of a uniform infill ratio, designers may employ functionally graded infill, where high infill percentages are used in high-stress areas, and lower infill rates are applied to low-stress areas. Additionally, topology optimization methods can be utilized to optimize infill ratio distribution. This topic will be discussed in detail in Chapter 6.

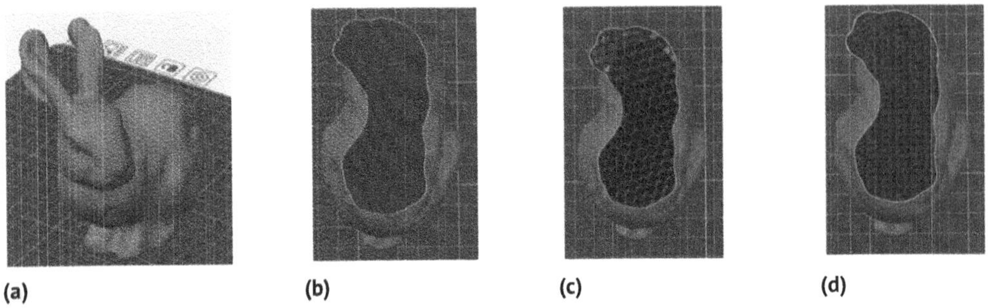

Figure 5.3 Different types of infills used to save printing time and material consumption. (a) Model of Stanford bunny volume 279629 mm^3, (b) 100% rectilinear infill–mass 339.45 g and printing time 7 h 37 min, (c) 15% honeycomb infill–mass 89.83 g and printing time 3 h 12 min, (d) 15% gyroid infill–mass 74.96 g and printing time 2 h 41 min.

Even though hollowing the part with porous infill can significantly decrease printing time and cost, it also presents some constraints, especially for AM processes that form geometry inside a bed or vat of raw materials, such as PBF and vat photopolymerization (VP). In these processes, removing extra raw materials, such as unmelted powders or residual resins, from inside closed structures remains a challenging task. Therefore, escape holes need to be designed in these cases.

For sheet lamination and material jetting processes, even with escape holes, it is still impossible to remove the extra unused materials inside the porous internal structures. Consequently, porous infill structures are seldom used for these two processes as well.

Based on the constraints of manufacturing cost and time, the following design guideline is summarized:

Design Guideline 5-3: During the design for the AM process, the part needs to be checked if it can be hollowed, or the mass of the part can be reduced to further reduce its manufacturing cost and time.

Previous hollowing or infill approach is one typical way to implement Design Guideline 5-3. Other optimization approaches that can further reduce the weight of designed components while not compromising their mechanical properties will be discussed in Chapter 6.

5.1.3.1 General Constraints 4: Support Structure

Support structures, as shown in the grey area of Figure 5.4, are essential in most AM processes, including ME, material jetting, Vat Photopolymerization (VP), and PBF for metals. Although these AM processes require support structures to successfully print overhangs, each uses them in slightly different ways. The specific principles governing the use of support structures will be discussed in detail in the next two subsections.

In contrast, a few AM processes, such as PBF for polymers and binder jetting, do not require support structures. Therefore, in these specific processes, minimizing support usage is unnecessary.

In general, the fabrication of support structures not only increases processing time and material usage but can also impact the quality of printed surfaces. Therefore, minimizing the use of support structures is essential in the design of AM parts. There are two general approaches to achieving this goal.

First, designers can optimize the printing orientation. For instance, in Figure 5.4, if the part is rotated 90 degrees counterclockwise, no support structures are needed. As a result, considerable

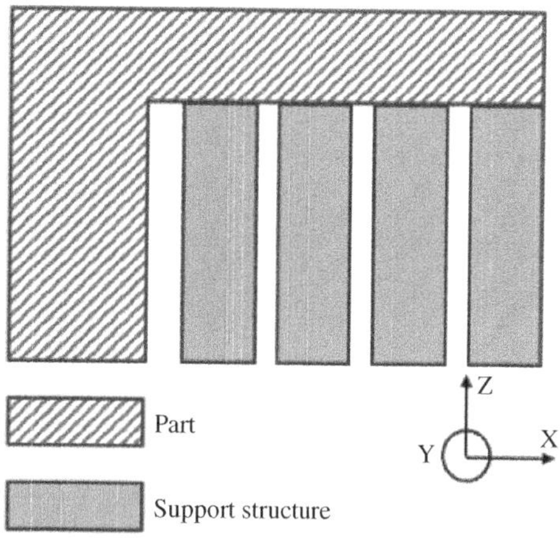

Figure 5.4 Demonstration of support structures.

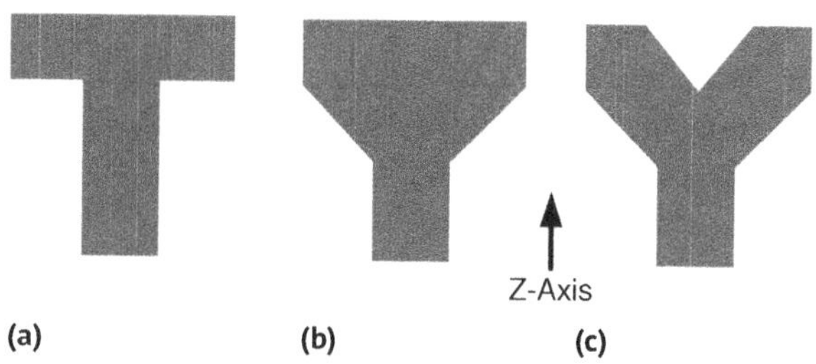

Figure 5.5 (a–c) Demonstration of reduced support-structure usage through overhang shape modification.

research focuses on optimizing printing orientation to reduce support usage. However, there are situations where using support structures is unavoidable. In some cases, no orientation achieves zero support, or the optimal orientation for minimal support may conflict with other criteria, such as extending printing time or destabilizing the part on the platform.

For these reasons, it remains essential during the design phase to minimize support structures whenever possible, provided this does not compromise the part's functional performance. This consideration is reflected in the following design guidelines.

Design Guideline 5-4: Once the printing orientation is determined, it is essential to check if the overhang surfaces can be modified to avoid using support structures.

Design Guideline 5-4 can be illustrated by the example shown in Figure 5.5. In this case, the original design is depicted in Figure 5.5a. Support structures are needed for this part due to the overhang regions on both sides. To avoid using supports, self-supporting surfaces can be added to the bottom of the overhang regions, as shown in Figure 5.5b. This modification allows the part to be printed successfully without support structures, significantly reducing the time required for their removal after the AM process. However, compared to Figure 5.5a, the weight of the modified design has increased.

To further reduce the weight, the part can be modified into a "Y" shape by removing material from the top center, as shown in Figure 5.5c. The final "Y" shaped part not only avoids increasing weight but can also be printed without support structures. It should be noted that this modification is only feasible if the downward-facing surface of the overhang region in part A is not a functional surface

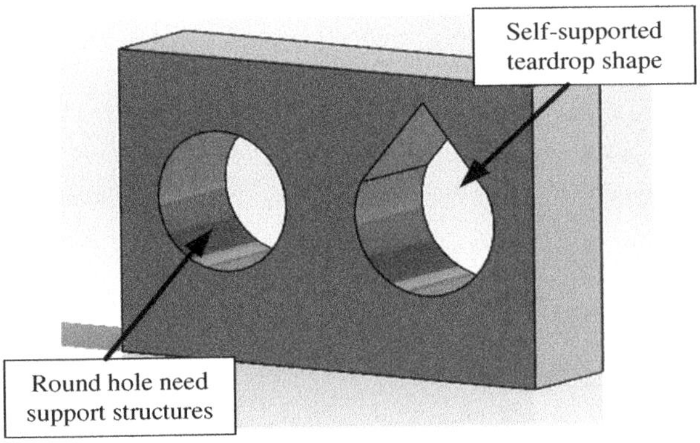

Figure 5.6 Teardrop shape hole can reduce the usage of support structure.

critical for assembly. Therefore, in some scenarios, the use of support structures is still inevitable. In such cases, ease of support structure removal should be considered during the design process.

Another example illustrating Design Guideline 5-4 is shown in Figure 5.6. Here, a typical horizontal hole needs to be printed, as shown in Figure 5.6. If the hole diameter exceeds a certain value, which will be discussed in the next subsection, the top part of this cylindrical surface cannot be successfully printed due to its overhang angle being smaller than the threshold for the selected AM process. In this case, a teardrop-shaped hole is redesigned, as shown in Figure 5.6. In the redesigned hole, the top part of the horizontal hole is replaced with overhang surfaces that can be printed without support structures.

5.1.3.2 General Constraints 5: Multi-material Complexity

One of the beauties of AM compared to conventional manufacturing processes is the enabling fabrication of multiple materials in a single part and the fabrication of functional gradient materials. This provides unprecedented design freedom for designers to further improve the performance of parts. However, it should be noted that there are still certain limitations that designers need to consider. These limitations can be summarized below:

- Materials compatibility: Different materials have different coefficients of thermal expansion, which may lead to warping or delamination due to large internal thermal stress.
- Adhesion or interface of multi-materials: Ensuring good adhesion or connection between different materials for the structural integrity of the printing. For the metal AM process, intermetallic compounds around the boundary between two materials may lead to brittle failure of printed structures or parts.
- Recycling: Recycling multi-materials parts is also very challenging. From a recycling point of view, designers should try to minimize the usage of multi-material parts.

Thus, during the design of the multi-materials part of the AM process, the general limitations summarized above should be considered.

5.2 Design Guidelines for Polymeric AM Process

Several AM processes can fabricate polymer materials, including ME, material jetting, binder jetting, vat photopolymerization, and polymer-based PBF. This subsection will provide detailed design guidelines for each of these AM processes.

5.2.1 Material Extrusion

The ME process, widely known as Fused Filament Fabrication (FFF) or Fused Deposition Modeling (FDM), fabricates parts by extruding material through a nozzle in successive layers, as shown in Figure 5.7. This is one of the most affordable AM technologies, capable of printing a wide range of polymeric materials, including thermoplastics like PLA and ABS, composite materials with both short and continuous fiber reinforcements, and even metal or ceramic parts. In this chapter, we will focus on its capabilities for fabricating polymer-based parts. Specifically, this subsection will discuss design constraints from two perspectives: material and geometry.

5.2.1.1 Design Consideration for Materials

The ME process can fabricate a wide range of thermoplastic materials. To assist designers in quickly selecting materials for their designs, Table 5.3 is provided comparing commonly used materials in this process. It should be noted that specific material properties are not included in this table, as they can vary significantly depending on factors such as printing orientation and process parameters. A detailed discussion of the specific property values and the influence of process parameters is available in the existing review [3].

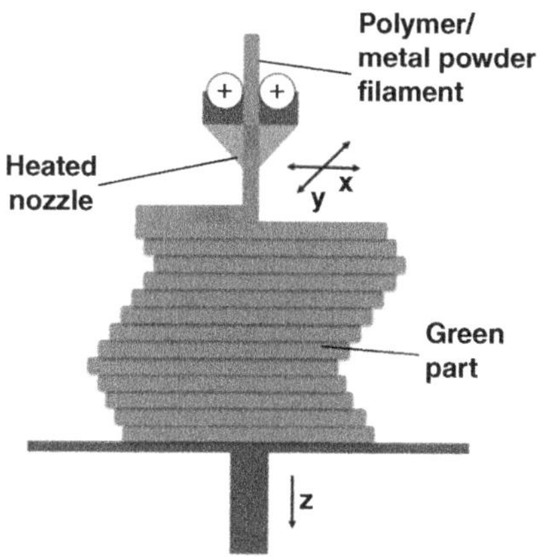

Figure 5.7 Working principle of material extrusion (ME) process. *Source*: [2]/Carlo Burkhardt HSPF/Wikimedia Commons/CC BY SA 4.0.

Table 5.3 Typical materials for material extrusion (ME) process.

Material name	Stiffness	Strength	Elongation	Ease of printing	Cost
PLA	High	Moderate	Low	Easy	Low
ABS	Moderate	High	Moderate	Moderate	Low
PETG	Moderate	High	High	Easy	Low
Nylon	Low	High	High	Moderate	Moderate
TPU	Low	Moderate	Very high	Difficult	Moderate
PEEK	High	Very high	Moderate	Difficult	Very high
PC	Moderate	High	Moderate	Moderate	Moderate
Carbon reinforced PLA	Very high	High	Moderate	Moderate	High

Compared to thermoplastic materials fabricated through other manufacturing processes, such as injection molding, samples made via ME exhibit high anisotropy. Table 5.4 and Figure 5.8 illustrate the mechanical properties of PLA in different directions post-printing. The primary cause of this anisotropy is shown in Figure 5.9a: interlayer bonding is significantly weaker than intralayer bonding. Additionally, air gaps between layers further reduce ultimate strength and elongation in the vertical direction.

When a load is applied on the XY plane, the situation becomes more complex due to the influence of the raster angle on mechanical properties. Generally, if the loading direction aligns with the deposition path (Figure 5.9b), the sample demonstrates the highest tensile strength. This strength decreases

Table 5.4 Mechanical properties of ME-fabricated PLA along [4]/Springer Nature/CC BY 4.0.

	Elastic modulus/MPa	Ultimate tensile strength/MPa	Elongation/%
Horizontal	2606	57.15	4.36
Side	2438	55.83	4.13
Vertical	2380	24.22	1.02

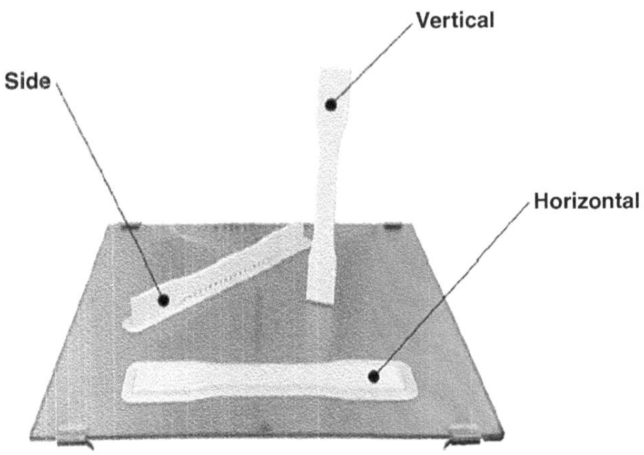

Figure 5.8 Different printing orientation for a standard tensile test coupon. *Source*: [4]/Springer Nature/CC BY 4.0.

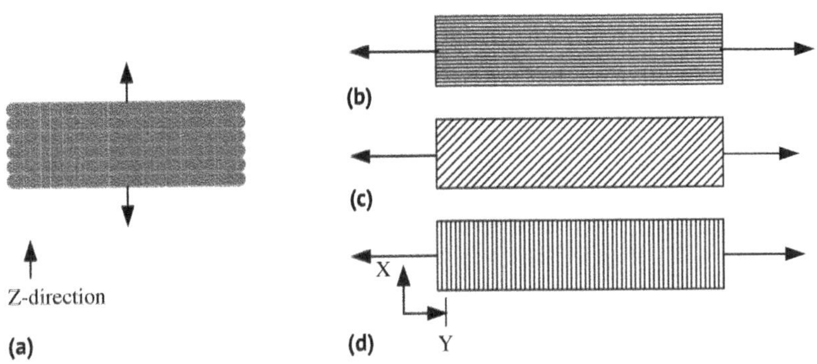

Figure 5.9 (a–d) Explaining the reason for anisotropic material properties fabricated by ME process.

as the angle between the loading direction and the path increases (Figure 5.9c), with the weakest strength occurring when the loading direction is perpendicular to the deposition path (Figure 5.9d).

Because of these anisotropic properties, designers must consider load conditions when selecting the printing orientation. It is also essential to verify strength constraints after choosing the orientation to prevent unexpected part failures due to anisotropy.

The ME process is particularly suited for multi-material printing, as it can utilize either a single nozzle to print multiple materials sequentially or multiple nozzles to print different materials simultaneously during part fabrication. This capability makes ME ideal for AM of composite materials and structures with functionally graded materials, where complex material property distributions are required. Additionally, water-soluble support materials can be used, simplifying the support removal process.

5.2.1.2 Design Considerations from Geometric Constraints

The design constraints for typical geometric features in the ME process are summarized in Table 5.5. It should be noted that the typical constraint values listed in this table are based on the most common ME machines using materials such as PLA and a 0.4 mm nozzle. These values can vary for different materials, machines, and process parameters. For example, the quality of overhangs can be optimized by adjusting the process parameters [5]. In such cases, the printable bridge length can be extended compared to the values provided below.

Besides process parameters, the geometric dimensioning and tolerancing (GD&T) requirements on these geometric features will also impact the threshold values for manufacturability. For instance, if there are high surface roughness requirements or tight tolerances on overhang

Table 5.5 Constraints on typical design features for the ME process.

Feature name	Illustration	Description	Typical value of constraints
Overhang without support	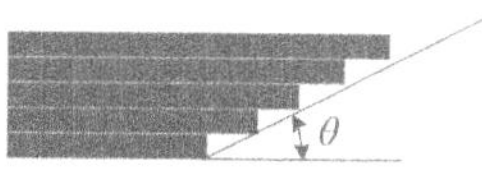	Support structures also need to overhang surfaces if the overhang angle θ is smaller than a certain value.	$\theta > 19°$
Bridge structure without support	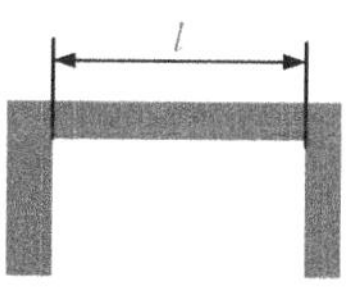	The bridge structure can be printed without support. The thicker the bridge is, the shorter it can be printed.	$l \leq 10\,\text{mm}$
Vertical pin		Due to the nozzle diameter, the diameter ϕ of a vertical pin should be larger than a threshold.	$\phi \geq 0.8\,\text{mm}$
Unsupported wall	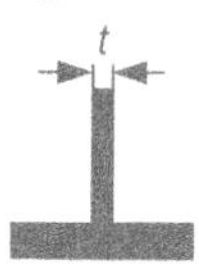	Due to the nozzle diameter as well as the stability of the printed feature, the minimal thickness of the wall should be larger or equal to t.	$t \geq 2\,\text{mm}$

(Continued)

Table 5.5 (Continued)

Feature name	Illustration	Description	Typical value of constraints
Embossed and engraved details	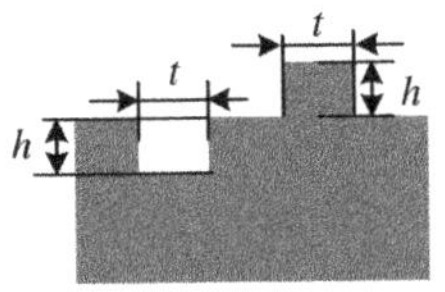	To make embossed and engraved features recognizable, the width *t* and depth *h* should be larger than certain thresholds, respectively.	$t > 0.6\,\text{mm}$ $h > 2\,\text{mm}$
Holes	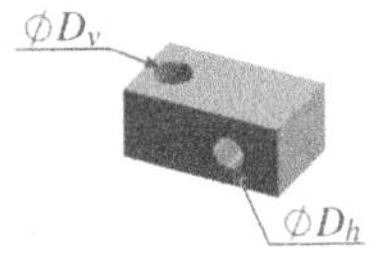	Due to the resolution of the manufacturing process, D_v diameter of the vertical hole should be larger than a certain threshold. Due to the overhang surfaces, D_h the diameter of the horizontal hole should be smaller than a certain value.	$D_v > 1.5\,\text{mm}$ $D_h < 10\,\text{mm}$
Horizontal corner	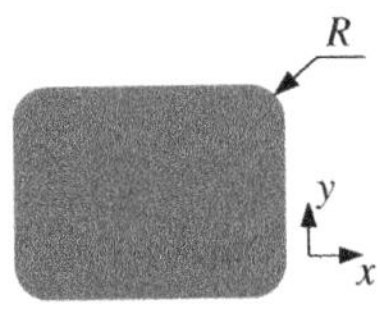	Due to the nozzle diameter, it is impossible to print sharp horizontal corners. The radius of horizontal corner *R* should be larger than a certain value.	$R > 0.4$ mm
Clearance between two moving components	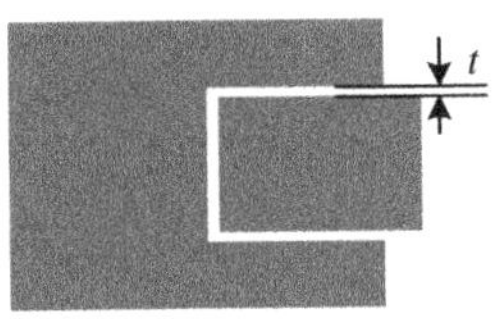	The clearance *t* should be designed to make sure two parts printed together can move freely.	$t > 0.5$ mm

surfaces, the minimum printable overhang angles θ might be larger than the 19° listed in Table 5.5. This rule also applies to other features such as vertical pins. When the pin has a strict tolerance requirement, the minimum diameter it can print is much larger than the threshold values listed in Table 5.5.

In addition to manufacturability constraints, several design guidelines can further improve the quality and functional performance of printed parts. The first guideline pertains to corner design, summarized as follows:

Design Guideline 5-5: Fillets are recommended to be used for the ME process, which can reduce stress concentration.

As illustrated in Figure 5.10, adding fillets to a design can help relieve stress concentrations at sharp corners or edges. This approach is especially important for edges or corners formed between vertical and horizontal surfaces, as shown in Figure 5.10. Weak material performance along the z-axis is a primary factor contributing to the failure of parts made via ME. By incorporating fillets, stress concentrations are reduced, which in turn enhances the overall strength of the design.

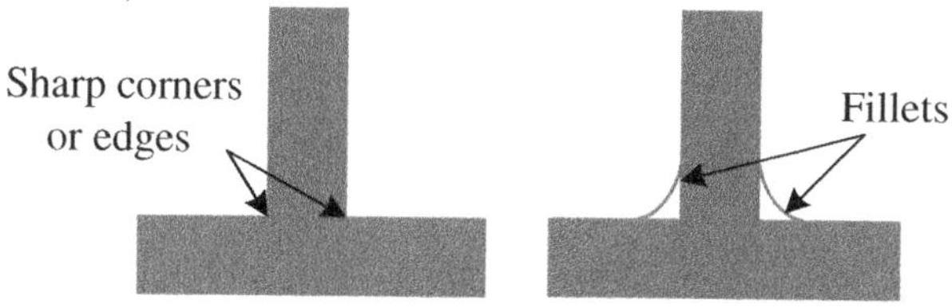

Figure 5.10 Illustration of adding fillets to relieve the stress concentrations.

Design Guideline 5-6: Once the printing orientation is determined, avoid the large cross-section area along the XY plane.

Once the printing orientation is determined, designers should check if the area of each printing layer can be further reduced. One typical method is to use infill structures, as discussed above. A large cross-sectional area may not only increase the printing time but, more importantly, lead to an uneven temperature distribution in the part after printing, causing warping.

5.2.2 Vat Photopolymerization

VP is also known as the SLA (stereolithography) process, which cures the liquid photopolymer resin layer by layer or volumetrically [6] using a light source. Like the ME process, this subsection will be divided into two parts: material considerations and geometric considerations.

5.2.2.1 Design Consideration for Materials

The liquid photopolymer resin used in the VP process is a chemical compound composed of monomers, oligomers, photoinitiators, and additives. By adjusting the ratio of these basic elements, materials for VP processes can be tailored to different properties. For example, tough resin, known for its durability and high-impact resistance, is ideal for snap-fit parts or applications requiring high-impact resistance [7]. For applications using the VP process to create investment casting patterns, castable resin is suitable as it can be burned out cleanly, providing high detail and a smooth finish [8]. In biomedical applications, dental resin is commercially available for dental models, surgical guides, and orthodontic devices [9]. Additionally, users can mix ceramic powders into the resin to achieve ceramic printing, although post-processing is required in this case [10]. In summary, VP processes provide quite high design freedom for material choice. Designers need to select a specific type of resin based on the requirements of their applications.

Unlike the ME process, the material properties of the VP process are generally isotropic, achieving the same elastic modulus and strength in the z-direction as in the X and Y directions. This makes the VP process especially ideal for applications where material isotropic properties are critical.

5.2.2.2 Design Consideration for Geometric Features

Like the ME process, there are still a few constraints that designers need to know during the design of structures fabricated by the VP process. Those geometric constraints are summarized in Table 5.6.

Table 5.6 Constraints on typical design features for VP processes.

Feature name	Illustration	Description	Typical value of constraints
Overhang without support	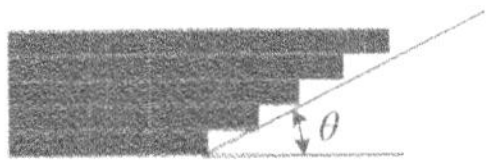	The surface of the overhang structure may sag due to gravitational force. A low overhang angle θ can result in poor and deformed surfaces.	$\theta > 45°$
Bridge structure without support	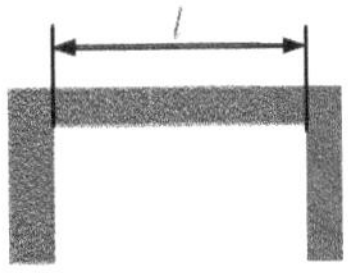	The bridge structure can be printed without support. Its length l should be smaller or equal to a threshold value.	$l \leq 20$ mm
Vertical pin		Pins are tall thin features with circular cross-sectional areas. If they are too thin, there is a risk of breaking.	$\phi \geq 0.8$ mm
Unsupported thin wall	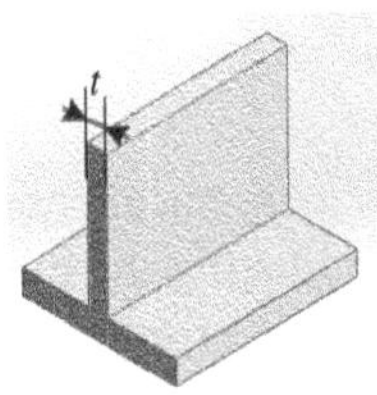	In the VP process, the unsupported thin wall is at high risk of warping or breaking off. To avoid that, the wall thickness t should be larger than a certain value.	$t \geq 0.8$ mm
Supported thin wall	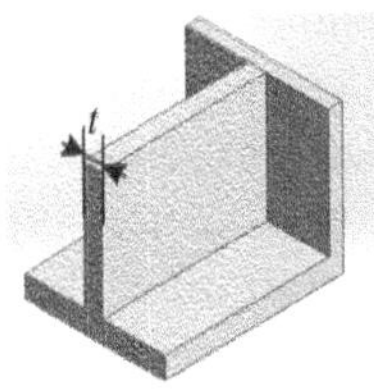	In the VP process, the supported wall has higher stiffness and strength and a lower risk of warping and breaking. Thus, the thinner it can be printed.	$t \geq 0.5$ mm
Engraved and embossed features	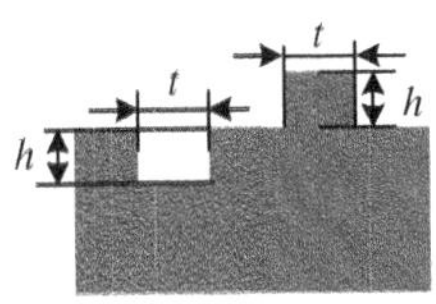	If those features are too small, there is a risk of fusing with the rest of the model during the printing. It is suggested to be larger than the given values.	$t > 0.4$ mm $h > 0.4$ mm
Holes	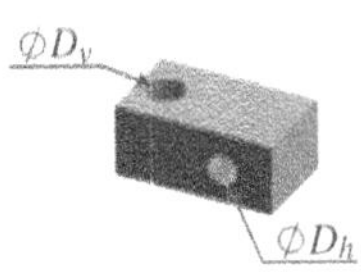	If the diameter of the hole is less than a certain value, it may close off during the fabrication.	$D_v > 0.8$ mm $D_h > 0.8$ mm

Feature name	Illustration	Description	Typical value of constraints
Clearance between two moving components	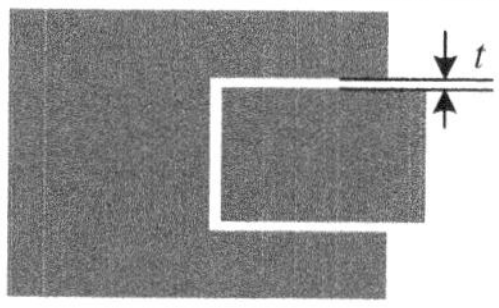	The clearance t should be designed to make sure two parts printed together can move freely.	$t > 0.5$ mm
Hollow parts	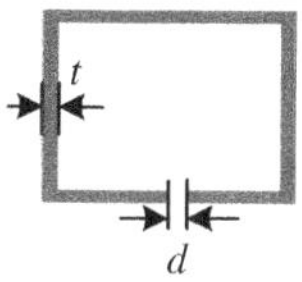	To make sure the uncured resin can flow out of the hollow parts, the drainage holes whose diameter d should be larger than a certain value. In addition, the thickness of wall t should also be larger than a given value.	$d \geq 3.5\,\text{mm}$ $t \geq 2\,\text{mm}$

Like ME processes, the constraint values listed in Table 5.6 depend on factors such as material type, process parameters, and machine configurations. Generally, VP processes can be categorized into two distinct machine configurations: bottom-up and top-down. In the bottom-up process, curing occurs at the bottom of the vat, where a peeling force is applied as the platform lifts. This peeling force increases the risk of breakage for delicate features, like thin walls or pins, during printing. In contrast, the top-down process cures resin at the surface, eliminating the peeling force, which makes it ideal for printing fine or small features. Consequently, threshold values vary between machines with these two configurations. To obtain accurate data for these constraint values, printing benchmark parts is recommended.

5.2.3 Selective Laser Sintering

SLS is a type of PBF process that uses a laser to melt or sinter thermoplastic particles layer by layer, forming a three-dimensional part. Currently, polyamide (PA), especially PA 12, is the most used material for the SLS process due to its excellent mechanical properties and good processability.

Compared to conventionally manufactured PA 12, SLS-fabricated PA 12 can achieve a similar elastic modulus of around 1600 MPa [11]. More importantly, its values along different axes are almost the same, demonstrating isotropic properties. This isotropy also applies to tensile strength, with only a very slight decrease along the Z-axis (45 MPa) compared to the X or Y axes (48 MPa) [12]. These values are close to those of conventionally manufactured PA 12. "The most notable difference between SLS-fabricated PA 12 and conventionally produced PA 12 lies in elongation." SLS-fabricated PA 12 typically exhibits approximately 20% elongation in the X and Y directions and about 8% in the Z direction [12]. This is much lower compared to conventionally manufactured PA 12. Therefore, SLS-fabricated PA 12 parts are not suitable for applications requiring high elongation. To further enhance the mechanical properties of PA 12, short carbon fiber can be filled inside the PA 12 powders. Both stiffness and strength can be enhanced by using carbon fibred-reinforced PA 12. Besides PA 12, thermoplastic polyurethane (TPU) is another widely used material for SLS. TPU provides relatively high elongation with lower stiffness.

In terms of geometric design considerations, one of the major advantages of SLS printing is that it does not require support structures, as the unmelted powder serves as a natural support. This makes the SLS process well-suited for fabricating complex geometries that are challenging to produce with other polymer-based AM methods due to the difficulties associated with support removal. However, SLS does have certain geometric limitations, which are summarized in Table 5.7.

Table 5.7 Constraints on typical design features for the SLS process.

Feature name	Illustration	Description	Typical value of constraints
Vertical pin		Due to the melting pool size as well as the heat effect zone, semi-melted powders will be attached around the surface. The diameter of pin ϕ should be larger than a certain value.	$\phi \geq 1$ mm
Wall thickness	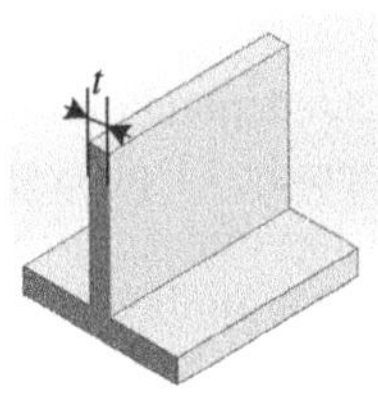	To ensure a successful fabrication, the wall thickness t should be larger than a certain value.	$t \geq 0.8$ mm
Engraved and embossed features	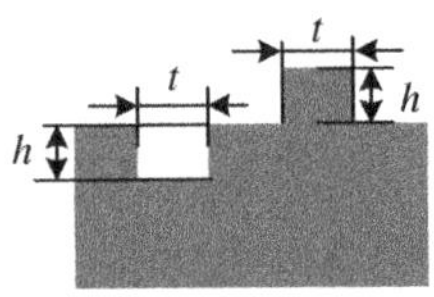	If those features are too small, there is a risk of fusing with the rest of the model during the laser sintering process. The suggested width t and height h of those features should be larger than a given value.	$t > 1$ mm $h > 1$ mm
Holes	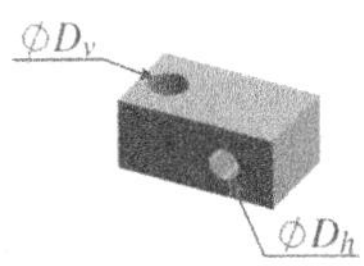	If the diameter of the hole is less than a certain value, the powder inside the hole might be difficult to remove.	$D_v > 1.5$ mm $D_h > 1.5$ mm
Clearance between two moving components	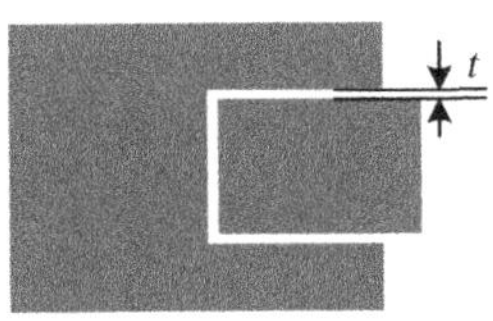	The clearance t should be designed to make sure two parts printed together can move freely.	$t > 0.5$ mm
Hollow parts	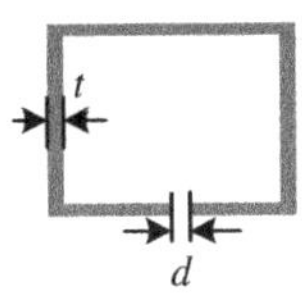	To make sure the un-sintered powders can flow out of the hollow parts, the drainage holes whose diameter d should be larger than a certain value. In addition, the thickness of wall t is also suggested to be designed larger than a given value.	$d \geq 3.5$ mm $t \geq 1$ mm

In summary, SLS offers significant design freedom, allowing parts to be printed without the need for support structures. However, it is important to note that there are still some limitations to the SLS process. First, small feature details can be challenging to print because radiant heat can fuse unsintered powder around these features. Additionally, shrinkage and warping can occur during the printing process, which can significantly impact accuracy and may even lead to cracks. To mitigate this, most SLS machines can compensate for shrinkage during the process planning stage. For warping, it is advisable to avoid parts with large cross-sectional areas. Finally, powder removal is another challenge in the SLS process, requiring the design of drainage holes to facilitate the removal of unsintered powder.

5.2.4 Material Jetting

Material jetting is another popular AM process for fabricating polymeric materials. It operates similarly to a two-dimensional inkjet printer, depositing material onto the build platform layer by layer. Compared to other AM processes for polymers, material jetting is ideal for creating parts with fine features due to the small droplet size, approximately 16 μm. More importantly, most material jetting machines support multi-material fabrication and can control material placement at the pixel level. This enables the direct printing of full-color parts by mixing polymer resins in various color ratios. This capability is particularly valuable for parts used in demonstrations and educational purposes, such as medical phantoms for training.

Like VP process, most material jetting processes also use photopolymers. By controlling its chemical composition, the cured resin can exhibit different mechanical properties. They can be either soft, rubber-like materials with high elongation or rigid ABS-like polymers for load-bearing functions. It can also digitally combine soft and hard materials to achieve functionally graded mechanical properties, which are hard to achieve by other AM processes [13, 14].

Like many other AM processes for polymeric materials, the material jetting process also requires the usage of support structures. Since it can print multi-materials, the water-soluble material is widely used for the material jetting process, which makes the removal of supporting materials much easier compared to other polymeric AM processes such as VP.

Typical applications of the material jetting process include full-color models, injection mould, and its prototype, enclosure as well as a physical model to show FEA results. Even if it can achieve small resolution and enable the printing of small features, there are still geometric constraints that designers need to consider. They are summarized in Table 5.8.

5.3 Design Guidelines for Metal AM Process

Compared to polymer AM processes, there are fewer types of metallic AM processes. Broadly, metallic AM can be classified into two main categories: PBF and DED. The design guidelines and considerations for each of these technologies will be discussed separately.

5.3.1 Powder Bed Fusion

The PBF process is a type of AM technology that creates complex parts by fusing the powders in the powder bed layer by layer. It can be used to fabricate both metallic and polymeric materials. The PBF process for polymeric materials is usually called SLS, which has already been introduced

Table 5.8 Constraints on typical design features for material jetting process.

Feature name	Illustration	Description	Typical value of constraints
Vertical pin		Pins with a diameter (ϕ) smaller than a certain value are prone to breaking during the removal of support structures.	$\phi \geq 0.5$ mm
Thin wall	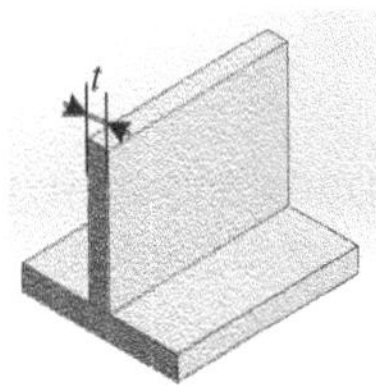	Wall thickness t smaller than a given value can be easily broken during the removal of support structures.	$t \geq 0.5$ mm
Engraved and embossed features	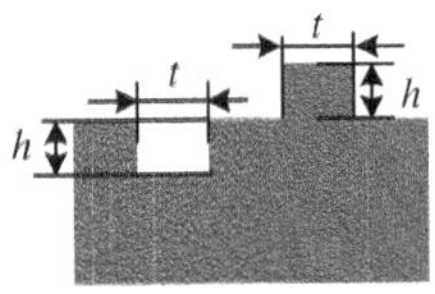	If those features are too small, it may be difficult to remove the support structures around them.	$t > 0.5$ mm $h > 0.5$ mm
Holes	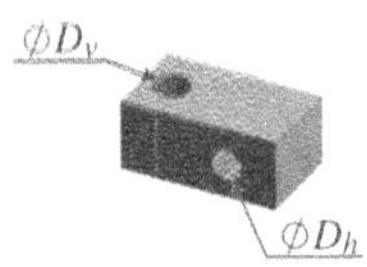	Small holes have issues with removal support structures and lose accuracy. Vertical holes have better accuracy compared to horizontal holes.	$D_v > 0.5$ mm $D_h > 0.5$ mm
Clearance between two moving components	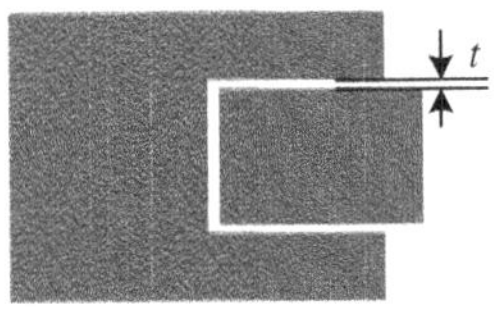	The clearance t should be designed to make sure two parts printed together can move freely.	$t > 0.15$ mm
Hollow parts	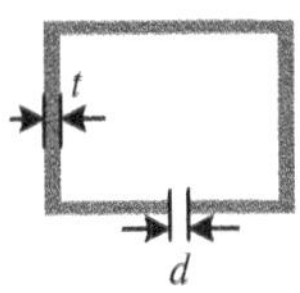	Try to avoid the design of hollow parts since the supports need to be filled in the hollow parts, and they are difficult to remove.	N.A

in Section 5.2.3. In this section, we are going to discuss the design consideration of PBF processes for metals. These PBF processes can be further divided into two categories based on the energy source they use for manufacturing. They are laser-powder bed fusion (L-PBF) and electron beam melting (EBM). Compared with the EBM process, the L-PBF process can generally provide a better surface finish and achieve better resolution. However, EBM can achieve larger melt pools due to higher energy input with less warping and thermal stress by preheating powder beds. It is generally ideal for printing large components. The type of design constraints for these two processes are almost the same. Thus, they will discuss together in the following sections.

Table 5.9 Tensile properties of Ti6Al4V fabricated by L-PBF and conventional manufacturing processes.

Alloys–process	Machine	Orientation	UTS (MPa)	YS (MPa)	Elongation (%)
Ti6Al4V-EBM [15]	Arcam A1	XY	833 ± 22	783 ± 15	2.7 ± 0.4
Ti6Al4V-EBM [15]	Arcam A1	Z	851 ± 19	812 ± 12	3.6 ± 0.9
Ti6Al4V-EBM [16]	Arcam S400	XY	917 ± 30.53	844 ± 21.6	8.8± 1.42
Ti6Al4V-EBM [16]	Arcam S400	Z	842 ± 13.84	782 ± 5.1	9.9 ± 1.02
Ti6Al4V-L-PBF [17]	EOS M270	XY	1269 ± 9	1195 ± 19	5 ± 0.5
Ti6Al4V-L-PBF [17]	EOS M270	Z	1219 ± 20	1143 ± 30	4.89 ± 0.6
Ti6Al4V-Wrought [18]	NA	Longitudinal	948	994	21
Ti6Al4V-Wrought [18]	NA	Transverse	962	1008	19

During the design stage, the first thing that needs to be considered is the mechanical properties of materials. The mechanical properties of metals fabricated by AM may be different compared to the same materials fabricated by conventional manufacturing processes such as casting or forging. This is due to the different microstructures from the different cooling speeds under different manufacturing processes. For example, typical mechanical properties of Ti6Al4V fabricated by different manufacturing processes are provided in Table 5.9. In general, the L-PBF process can achieve better ultimate tensile stress (UTS) as well as yield stress (YS) compared to other manufacturing processes. However, the elongation of AM-fabricated materials is lower than that manufactured by the wrought process. Another thing that should be noticed is that even for the same process, different machines may lead to different mechanical properties. The mechanical properties are sensitive to multiple factors from a machine such as the powdering spreading mechanism, protection gas flow, as well as process parameters used. All those factors will lead to the change of microstructures of fabricated materials and finally affect their mechanical properties. Thus, for loading critical parts, it is strongly suggested the test coupons need to be printed to get the mechanical properties of fabricated under certain fabrication scenarios.

In addition to the effects of process parameters, designers should also be aware that geometric factors can affect the microstructures of fabricated parts. For example, overhang regions are likely to overheat. These different thermal histories will lead to different microstructures, which will ultimately affect their mechanical properties.

Compared to many polymer-based AM processes such as ME, the PBF process can achieve nearly isotropic material properties, simplifying considerations around print orientation. Designers do not need to account for the impact of printing orientation on the part's mechanical performance. Instead, the focus should be on minimizing the use of support structures and enhancing print quality, including reducing warping caused by uneven thermal stress and addressing rough surfaces due to semi-melted powder attachments.

Besides material properties, there are also a few geometric feature-based constraints that designers need to consider during the design of components fabricated by the PBF process. They are summarized in Table 5.10.

It should be noted that the constraints summarized in Table 5.10 apply specifically to the L-PBF process. However, these types of constraints can also be applied to the EBM process, though the constraint values may require slight adjustments. Additionally, as with other AM processes, different materials and machines may have varying constraint values. Therefore, it is recommended to fabricate benchmark parts to assess the capabilities of each machine.

Table 5.10 Geometric design constraints for L-PBF process.

Feature name	Illustration	Description	Typical value of constraints
Vertical pin		The diameter ϕ of the thin vertical pin is limited by the melting pool size as well as the scanning strategy.	$\phi \geq 0.8$ mm
Thin wall	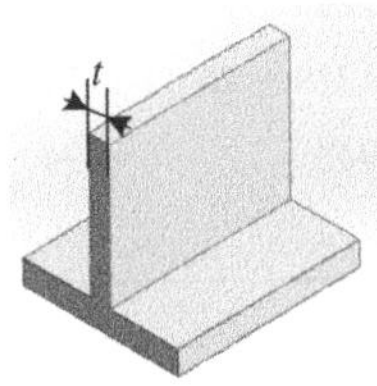	Thin walls are easy to be damaged due to warping.	$t \geq 0.4$ mm
Overhang without support	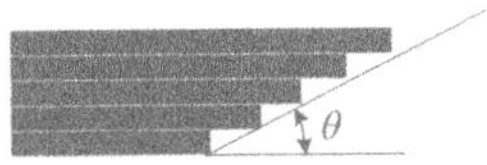	The surface of the overhang structure may sag due to gravitational force. A low overhang angle θ can result in poor and deformed surfaces.	$\theta > 45°$
Engraved and embossed features	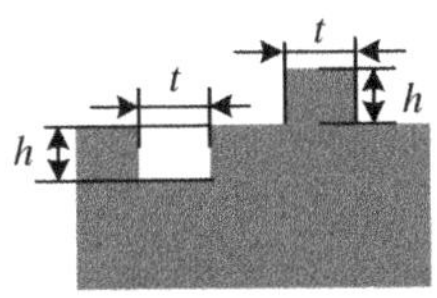	If those features are too small, it may be difficult to remove the support structures around them.	$t > 0.2$ mm $h > 0.2$ mm
Holes	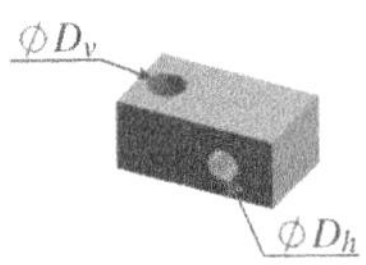	Small holes make it difficult to remove powders. For horizontal holes, if their diameter is larger than a certain value, they need to be printed with support structures.	$D_v > 0.6$ mm 0.6 mm $< D_h <$ 10 mm (without support)
Clearance between two moving components	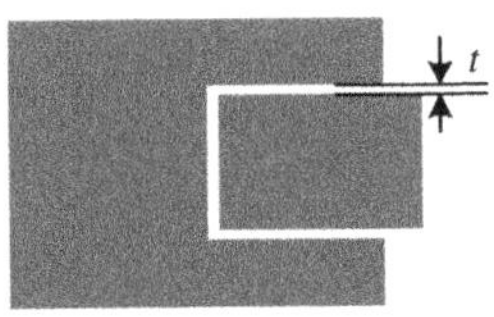	The clearance t should be designed to make sure two parts printed together can move freely.	$t > 0.2$ mm
Hollow parts	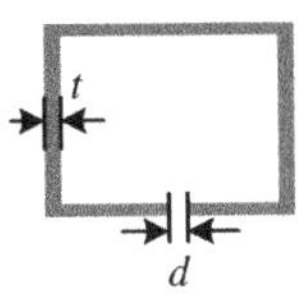	Holes are needed on hollowed metal parts so unmelted powder gets properly removed. The thin wall for the hollow part is likely to warp during the printing.	$d > 3$ mm $t > 0.8$ mm

5.3.2 Direct Energy Deposition

DED is another key AM process used for manufacturing metallic materials. It employs energy sources such as lasers, electron beams, electric arcs, or even plasma to melt metallic powders or wires during the fabrication process. The melted material is deposited onto a substrate layer by layer, ultimately shaping the desired part geometry. A comparison table shown in Table 5.11 is

Table 5.11 Comparison between the DED and PBF process.

Feature	Direct energy deposition	Powder bed fusion
Raw materials	Wire/Powders	Powders
Energy source	Laser, electron beam, plasma arc, electric arc	Laser or electron beam
Build size	Large	Small to medium
Part complexity	Low complexity	High complexity
Speed	Faster for larger parts	Slower due to precision
Surface finish	Rough, needs post-processing	Medium to fine, may still need post-processing
Cost	Lower comparing with PBF	High
Multi-material capability	Use multiple materials in the same build	Difficult to print multi-material part
Residual stress	High	Medium
Applications	Repair, large part near-net shape	Complex, high-value, precision parts

provided to help designers decide which metallic AM process is most suitable for their applications. A major advantage of the DED process is its ability to fabricate multi-material components within a single part, making it ideal for applications that require multiple metallic materials, such as rocket engines. Additionally, DED's capability to deposit material onto existing parts makes it suitable for repairing or remanufacturing. However, there are some limitations to the DED process. One significant drawback is the rough surface finish, which typically requires post-machining. To address this, hybrid manufacturing machines that combine DED with machining have been developed, allowing for the direct fabrication of parts with high geometric and dimensional accuracy. Another challenge is warpage caused by cyclic thermal stress during printing. Parts may require intermediate stress relief during deposition, or design compensation should be considered in advance.

When it comes to design for the DED process, the material characteristics of DED-processed metals need to be considered first. In most cases, metals processed via DED often exhibit large, columnar grains due to the localized and directional solidification during deposition. The rapid cooling rates can lead to a nonuniform microstructure, with significant anisotropy where the mechanical properties vary in different directions [19]. This can affect the mechanical properties of large parts. In addition, the DED process also tends to have higher porosity and greater chance of defects, and these defects act as stress concentrators, reducing the effective load-bearing capacity of materials. Thus, for those load-critical applications, post-processing such as hot isostatic pressing [19] is needed.

As to geometric constraints, DED process has its limitations on the fine features such as thin pins or holes with smaller diameters. The value of those constraints is different from different types of DED processes. In general, powder-based DED offers higher accuracy and finer detail compared to the wire-based DED process. However, the wire-based DED process has higher material and cost efficiency. The size limitation of fine features such as thin walls or pins it can print is determined by its bead width. Typically, the bead or track width of powder-based DED ranges from 0.5 to 2 mm while wire-based DED ranges from 1.5 to 5.0 mm. However, a thicker bead or track is also available for some special nozzle design. The minimum feature size should be larger than the thickness of a single bead or track.

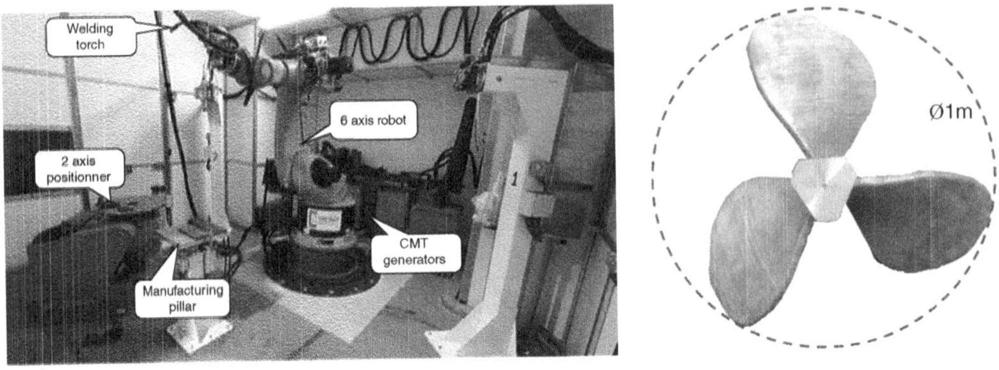

Figure 5.11 Multi-axis DED fabrication process. *Source*: [20]/MDPI/CC BY 4.0.

In addition to the limitations on fine features, another major design constraint that should be considered for the DED process is the overhang structures. Like the ME process, the DED process also needs support structures for the printing of overhang structures. However, the DED process cannot use the complex porous support that has been widely used for the PBF process due to its limitations on the fabrication of fine features. Solid support makes it much more difficult to remove after the printing. For the DED machine with a conventional 3-axis arrangement, the maximum achievable angle θ should be larger than 75°. This angle can be increased by using multi-axis arrangement including tilting the stage or deposition using multi-axis robotics arms or machines with complex path planning. For example, in Figure 5.11, a 6-axis arm is used with a 2-axis positioner. Based on a dedicated toolpath planning algorithm [20], it enables the printing of a propeller whose overhang angle is smaller than the general threshold of 75°.

Another important design consideration for the DED process is to consider the manufacturability of post-machining or manufacturing. Fixturing position or datum needs to be considered during the design of DED fabricated parts. By doing that, it can significantly reduce the cost and time for post-processing.

5.4 Design Guidelines for Other AM Processes

Besides those AM processes discussed above. Few AM processes are widely used for both polymeric and metallic materials. The design considerations for those AM processes will be summarized in this section.

5.4.1 Binder Jetting

The binder jetting process is a type of AM that selectively deposits a binding agent onto a powder bed to create solid parts layer by layer. Unlike other AM processes, binder jetting does not require high energy input and can be used with a wide range of materials, including metals, polymers, ceramics, and sand. However, for most materials produced through binder jetting, post-processing is required to achieve the desired material properties and final part quality. After printing, the fragile "green part," held together only by the binder, must undergo several post-processing steps such as curing, sintering, and infiltration. These steps improve the density and mechanical properties of the fabricated components but may also introduce side effects, including shrinkage, cracking, or oxidation if protective gas is not properly used.

Like other powder-bed-based processes, binder jetting has limitations when it comes to fabricating multi-material components within a single part. However, pigments can be added to the binders, enabling the printing of colorful prototypes. Another consideration for binder jetting materials is mechanical performance; parts produced with this process tend to exhibit lower strength, ductility, and fatigue resistance compared to those manufactured through traditional methods, such as casting or forging, and even some other AM processes [21]. This is largely due to the presence of porosity.

Regarding geometric constraints, specific limitations for various design features are summarized in Table 5.12. A major challenge in printing fine features with binder jetting arises during the depowdering process. Small features are prone to damage during powder removal, and this also applies to small holes where powder is difficult to clean. The deeper the hole, the larger its diameter should be designed to ensure successful depowdering.

Compared to other AM processes like ME, binder jetting does not require support structures, making it particularly suitable for fabricating complex structures such as porous lattices.

Table 5.12 Geometric design constraints for Binder Jetting process.

Feature name	Illustration	Description	Typical value of constraints
Vertical pin		The diameter ϕ of a thin vertical pin cannot be too small, otherwise it may be damaged during the depowdering stage	$\phi \geq 2$ mm
Thin wall	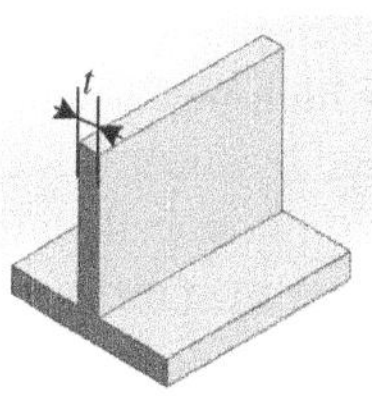	The recommended minimum wall thickness for parts produced via binder jetting is 2.0 mm. This ensures that the part can be removed from the powder and handled in the green state without being damaged	$t \geq 2$ mm
Engraved and embossed features	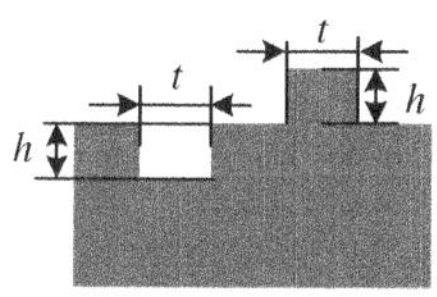	If those features are too small, it may be difficult to remove the support structures around them	$t > 0.5$ mm $h > 0.5$ mm
Holes	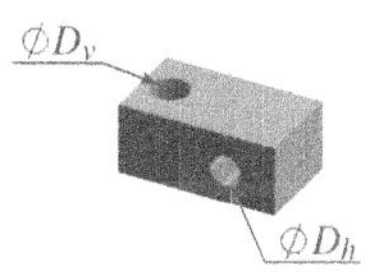	Small holes make it difficult to remove powders. Also, a through hole is suggested if possible to make it easy for depowdering	$D_v > 1.5$ mm $D_h > 1.5$ mm
Hollow parts	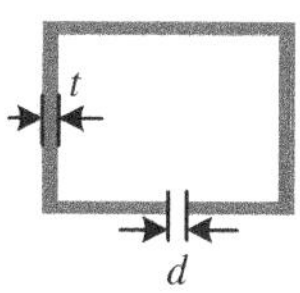	Holes are needed on hollowed parts so extra powder gets properly removed. The thin wall for the hollow part is likely to warp during the sintering	$d > 5$ mm $t > 2$ mm

However, designers should be aware that the sintering process can lead to shrinkage and distortion. Therefore, for applications with high GD&T requirements, post-machining may be necessary. To avoid cracking during the post-sintering process, it is recommended to incorporate a minimum fillet radius of 1 mm in the design of parts fabricated by binder jetting, as this minimizes stress concentration during sintering.

5.4.2 Sheet Lamination

Sheet lamination is an AM process that builds parts layer by layer by bonding sheets of material together. It is one of the least common AM processes available on the market. Any materials that can be made into sheets such as paper, plastic, or metals can be used for this type of AM process. After bonding, each layer is cut to the desired shape using a laser, knife, or other cutting tools. The excess material is removed after all layers are cut and bonded. Strictly speaking, sheet lamination can be considered as the combination of additive and subtractive manufacturing. Additive here means the materials are bonded layer by layer while subtractive refers to the cutting process. Compared to other AM processes such as ME, sheet lamination doesn't need extra support since the uncut materials play a role as support. Another major advantage is its cost-effectiveness, especially sheet lamination processes based on paper-based materials. In addition to that, the inks can be printed on paper, which enables the fabrication of low-cost colorful prototypes. Besides using inks, sheet lamination can also achieve multi-material printing by changing the material sheet during the manufacturing process. However, compared to other multi-material printing AM processes such as ME or jetting, sheet lamination can only achieve the material difference along the axis since the same layer should be always composed of a single material.

While sheet lamination offers unique capabilities, it also has notable limitations. This process typically does not provide the same level of mechanical strength or precision as PBF for metals or material jetting for polymers, making it less suitable for high-performance parts. Additionally, the surface quality in sheet lamination is largely determined by sheet thickness, which generally ranges from 0.05 to 1 mm. Thicker layers can result in a pronounced stair-step effect, affecting the surface roughness of the fabricated parts. More importantly, fabricating parts with internal cavities or hollow structures is challenging in sheet lamination due to the difficulty of removing excess material after cutting. Consequently, infill or hollow structures are not recommended for reducing the weight of parts produced by sheet lamination. During the design stage of parts fabricated by sheet lamination, designers first need to consider the properties of materials fabricated via the sheet lamination process. Compared to other typical AM processes, sheet lamination may produce parts with highly anisotropic material properties. Along the build direction, namely the z-axis, the strength of the material is mainly influenced by the strength of the adhesive or bonding mechanism between layers. Thus, the strength properties along this direction are much weaker than in other directions.

As to the constraints on the geometric features of the sheet lamination process, they can mainly be divided into categories. The first group of constraints mainly focuses on the fine features and precision it can achieve. Typically, the minimal feature size for a pin or thin wall that the sheet lamination process can achieve is determined by two major factors: cutting tool precision and the thickness of the sheet. For cutting tools, the cutting precision varies from different cutting tools. In general, for most laser-based cutting tools, the thickness of typical thin wall or vertical pin features should not be smaller than 0.1 or 0.2 mm. As to holes, their diameter should be larger than 0.3–0.5 mm depending on the precision that the cutting tool can achieve. Also, due to the size of cutting tools, the sharp horizontal corner is usually difficult to achieve as illustrated in Figure 5.12. Thus,

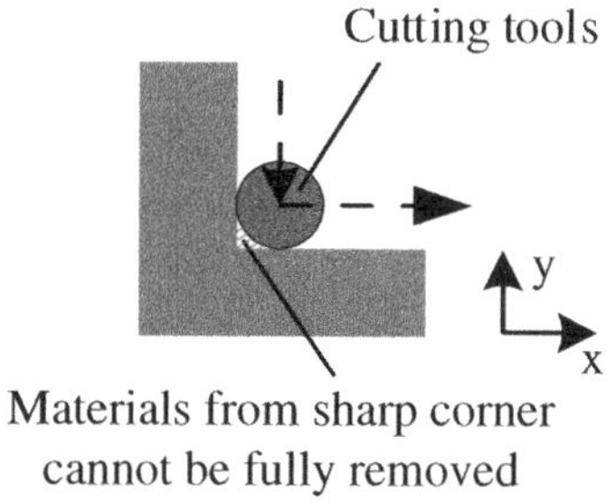

Figure 5.12 Avoid sharper corners in sheet lamination.

Figure 5.13 Removal of uncut materials of sheet lamination process.

a round corner is suggested to be designed to avoid those sharp horizontal corners. Besides the precision of cutting tools, the thickness of the sheet also limits the minimal feature size sheet lamination process can achieve. To ensure structural stability and avoid damage during the removal of uncutting materials, wall thickness or diameter of pins are generally suggested to be 2–3 times the sheet thickness.

The second group of constraints in sheet lamination arises from the removal of uncut material. Both small vertical and horizontal holes present challenges for material removal, particularly when they have a high aspect ratio (depth-to-diameter). Such features can be difficult to fabricate cleanly, so it is advisable to avoid them during the design stage. For hollow parts or enclosed cavities, a clear strategy for uncut material removal should be incorporated into the design. As illustrated in Figure 5.13, uncut material in design option (a) is significantly easier to remove than in option (b).

5.5 Conclusions

In summary, AM has significantly expanded the design space for parts with complex geometries and material compositions compared to conventional manufacturing methods. However, as discussed in this chapter, designers must still account for various constraints during the AM design process. These include limitations related to building volume, precision, support structures, and other process-specific factors. The values provided in this chapter serve as general guidelines, but specific constraints vary depending on the machine, process, and materials used. As such, benchmark testing remains essential for determining accurate parameters for different setups.

Importantly, addressing these constraints during the design phase can greatly reduce fabrication failures and help achieve the desired geometry and mechanical performance. Many of the constraints outlined are also orientation dependent. Since part orientation is typically determined during process planning—after the initial design—it is crucial to revisit the geometry using established guidelines at that stage. Multiple design iterations are often required to further optimize manufacturability.

References

1 Ali, M.H., et al., *Development of a large-scale multi-extrusion FDM printer, and its challenges*. International Journal of Lightweight Materials and Manufacture, 2023. **6**(2): pp. 198–213.
2 Commons, W, *File:Sinterbasiertes AM - Material Extrusion.png --- Wikimedia Commons{,} the free media repository*. 2023 [cited 2024 26-Sep-2024]; Available from: https://commons.wikimedia.org/w/index.php?title=File:Sinterbasiertes_AM_-_Material_Extrusion.png&oldid=723156640.
3 Goh, G.D., et al., *Process-structure-properties in polymer additive manufacturing via material extrusion: a review*. Critical Reviews in Solid State and Materials Sciences, 2020. **45**(2): pp. 113–133.
4 Morettini, G., et al., *Comprehensive characterization of mechanical and physical properties of PLA structures printed by FFF-3D-printing process in different directions*. Progress in Additive Manufacturing, 2022. **7**(5): pp. 1111–1122.
5 Dong, G., et al., *Optimizing process parameters of fused deposition modeling by Taguchi method for the fabrication of lattice structures*. Additive Manufacturing, 2018. **19**: pp. 62–72.
6 Stüwe, L., et al., *Continuous volumetric 3D printing: xolography in flow*. Advanced Materials, 2024. **36**(4): p. 2306716.
7 Javvaji, V.K., Angadi, S., and Piovesan, D., Optimizing material properties for 3D printing: a study on compressive strength of mixed clear and tough resins. in *ASME International Mechanical Engineering Congress and Exposition*. 2023, American Society of Mechanical Engineers.
8 Mukhtarkhanov, M., Perveen, A., and Talamona, D., *Application of stereolithography based 3D printing technology in investment casting*. Micromachines, 2020. **11**(10): p. 946.
9 Hata, K., et al., *Development of dental poly (methyl methacrylate)-based resin for stereolithography additive manufacturing*. Polymers, 2021. **13**(24): p. 4435.
10 Truxova, V., et al., *Ceramic 3D printing: comparison of SLA and DLP technologies*. MM Science Journal, 2020. **2020**: pp. 3905–3911.
11 Mehdipour, F., Gebhardt, U., and Kästner, M., *Anisotropic and rate-dependent mechanical properties of 3D printed polyamide 12 - a comparison between selective laser sintering and multi jet fusion*. Results in Materials, 2021. **11**: p. 100213.
12 Lammens, N., et al., *On the visco-elasto-plastic response of additively manufactured polyamide-12 (PA-12) through selective laser sintering*. Polymer Testing, 2017. **57**: pp. 149–155.
13 Salcedo, E., et al., *Simulation and validation of three dimension functionally graded materials by material jetting*. Additive Manufacturing, 2018. **22**: pp. 351–359.
14 Tang, Y., et al., *Universal material template for heterogeneous objects with applications to additive manufacturing*. Computer-Aided Design, 2020. **129**: p. 102929.
15 Edwards, P., O'conner, A., and Ramulu, M., *Electron beam additive manufacturing of titanium components: properties and performance*. Journal of Manufacturing Science and Engineering, 2013. **135**(6): p. 061016.
16 Khalid Rafi, H., et al., Mechanical property evaluation of Ti-6Al-4V parts made using electron beam melting. 2012 *International Solid Freeform Fabrication Symposium*, 2012. Austin Texas U.S.
17 Rafi, H., et al., *Microstructures and mechanical properties of Ti6Al4V parts fabricated by selective laser melting and electron beam melting*. Journal of Materials Engineering and Performance, 2013. **22**: pp. 3872–3883.

18 Shunmugavel, M., Polishetty, A., and Littlefair, G., *Microstructure and mechanical properties of wrought and additive manufactured Ti-6Al-4 V cylindrical bars*. Procedia Technology, 2015. **20**: pp. 231–236.

19 Tian, X., et al., *Isotropic and improved tensile properties of Ti-6Al-4V achieved by in-situ rolling in direct energy deposition*. Additive Manufacturing, 2021. **46**: p. 102151.

20 Rauch, M., Hascoet, J.-Y., and Querard, V., *A multiaxis tool path generation approach for thin wall structures made with WAAM*. Journal of Manufacturing and Materials Processing, 2021. **5**(4): p. 128.

21 Xu, M., et al., *Mechanical properties and microstructural characteristics of 316L stainless steel fabricated by laser powder bed fusion and binder jetting*. Journal of Materials Research and Technology, 2023. **24**: pp. 4427–4439.

6

Topology Optimization, Structural Optimization, and Multifunctional Optimization

This chapter lays the groundwork for design and optimization techniques in design for additive manufacturing (DfAM). It begins by discussing fundamental concepts in engineering optimization. Following this, structural and topology optimization for additive manufacturing (AM) is explained, with illustrative design cases for various AM processes. To support multifunctional design optimization, key concepts of multi-disciplinary design optimization are also introduced. The chapter concludes with a discussion of recent advancements in reliability-based design optimization (RBDO).

6.1 Basic Concepts of Engineering Optimization

Engineering optimization is highly relevant to DfAM. AM allows for complex geometries and innovative designs that are often impossible or impractical with traditional manufacturing methods. By incorporating optimization techniques, engineers can fully exploit the capabilities of AM to create lightweight structures, reduce material waste, and enhance functionality. Optimization in DfAM helps in determining the best part orientation, minimizing support structures, and ensuring that the final part meets performance requirements while considering manufacturing constraints. This integration of optimization with DfAM enables designers to push the boundaries of what is possible in product development, leading to more efficient and innovative solutions. To introduce the basics of engineering optimization approaches, the rest of this section is organized as follows. First, the basic formulation of optimization will be introduced. Then, different types of optimization problems will be reviewed. Lastly, the different optimization solvers will be discussed. Their advantages and disadvantages will be compared.

6.1.1 Formulation of Engineering Optimization

In general, optimization is a process aimed at finding the best values for one or more objective functions within a defined domain. This process, essentially a mathematical problem, is often referred to as mathematical optimization or mathematical programming. In engineering, optimization is a valuable tool to support decision-making and solve engineering problems. For instance, in structural design, optimization can be used to determine the optimal thickness of truss elements to achieve the desired stiffness with minimal weight. This approach is commonly known as engineering optimization.

The engineering optimization process begins with modeling, which involves identifying and mathematically describing the engineering problem. For any engineering problem intended to be addressed through mathematical optimization, it is essential to develop an optimization formulation. This formulation includes three fundamental components: the objective function, optimization or design variables, and constraint inequalities and equalities. These basic terms will be introduced in detail in the following sections of this subsection.

- **Objective function:** The objective function is a mathematical function representing the goal of an optimization problem. The optimal value of the objective function is determined by solving the optimization problem within the given constraints. This optimal value can either be a maximum or a minimum. For example, in structural design optimization, structural compliance can be set as the objective function to minimize it under a specified material mass constraint.

For a single engineering problem, there may be multiple ways to define the objective function. In structural design, we could alternatively define material mass as the objective function. Different objectives can lead to different optimal solutions, so domain knowledge from various engineering fields is necessary to choose the appropriate objective functions.

It should be noted that not all engineering optimization problems have an optimal solution. For those that do, there is a distinction between local and global optima. If the objective function is convex and the feasible region or design space forms a convex set, any local optimum is also the global optimum. However, non-convex problems are more challenging, as they may have multiple local optima, making it difficult to guarantee finding the global optimum.

In some engineering problems, multiple design objectives need to be achieved, which may lead to multiple objectives. In multi-objective optimization, it is important to note that a single solution may not exist that can simultaneously achieve optimal results for each objective. The objective functions may conflict, requiring trade-offs. Therefore, a solution to this type of optimization problem is called non-dominated or Pareto optimal, meaning that none of the objective functions can be improved without degrading at least one of the other objectives. There may be an infinite number of Pareto optimal solutions, so engineering optimization in this case can only provide a set of optimal solutions. Designers must then make a final decision based on domain knowledge to determine which solution will be used.

- **Optimization variables:** Optimization variables, or design variables, are the variables in the objective functions that can be adjusted to achieve the optimal solution. The set of all possible values for the optimization variables that satisfy the problem's constraints is called the design space or search space. For an optimization problem with n optimization variables, the design space is a subspace of an n-dimensional space. Generally, a higher-dimensional design space requires more computational effort to find the optimal solution.
- **Constraints**: In an optimization problem, a constraint is a condition that the solution must satisfy. Constraints define the boundaries or limitations within which the optimal solution must be found. They restrict the values that the optimization variables can take, shaping the feasible region where the optimization process occurs. Constraints can generally be classified into two types: equality and inequality. In an equality constraint, the function of the optimization variables must be equal to a specific value, while in inequality constraints, the function of the variables must be greater than or less than a specific value. Constraints can represent physical limitations, design requirements, budgetary limits, and safety criteria in engineering problems. Beyond these limitations, constraints can also describe physical laws, such as Hooke's law, that engineering problems must adhere to. The specific function used in the constraints is called the governing equation.

Based on the above discussion, the following standard formulation is summarized:

$$\begin{aligned} \min(\max) \quad & f_l(X), l = 1, 2, \ldots, p, p \geq 1 \\ & X = (x_1, x_2, \ldots, x_n) n \geq 1 \\ \text{subject to} \quad & g_i(X) \leq 0, i = 1, 2 \ldots, m, m \geq 0 \\ & h_j(X) = 0, j = 1, 2 \ldots, q, q \geq 0 \end{aligned} \tag{6.1}$$

where X is an n-dimensional design variable. $f_l(X)$ are objective functions. If $p = 1$, it is a single-objective optimization problem; otherwise, it is a multi-objective optimization problem. $g_i(X)$ and $h_j(X)$ are inequality and equality constraints, respectively. It should be noted that there is no special relationship between the number of design variables and the number of constraints. If $m, q = 0$, the optimization problem is called the unconstrained optimization problem.

In engineering optimization, the first step is to convert the engineering problem into the standard optimization problem shown in Formulation 1. Particularly, in many engineering design problems, state variables are dependent on the design variables and describe the system's behavior that will be used. For example, during structural design, state variables can be displacement or stress, and even temperature distribution. The relationship between state variables and design variables can be governed by physical laws or equations. These physical laws and equations can also be considered as equality constraints in the optimization problem. In addition to that, the objective function can also be defined based on the combination of state variables and design variables. In this case, Formulation 1 can be rewritten into:

$$\begin{aligned} \min(\max) \quad & f_l(X, Y(X)), l = 1, 2, \ldots, p, p \geq 1 \\ & X = (x_1, x_2, \ldots, x_n), p \geq 1 \\ & Y = (y_1, y_2, \ldots, y_o) o \geq 0 \\ \text{subject to} \quad & g_i(X, Y) \leq 0, i = 1, 2 \ldots, m, m \geq 0 \\ & h_j(X, Y) = 0, j = 1, 2 \ldots, q, q \geq 1 \end{aligned} \tag{6.2}$$

where X is the n-dimensional design variable and Y is the o-dimensional state variable that depends on design variable X. The relationship between X and Y can be described in the equality constraints which is usually called as governing equation. Numerical solutions such as the finite element approach can be used to solve those equations and obtain their explicit relationship.

6.1.2 Types of Engineering Optimization

After the modeling process, the next step is to solve the defined engineering optimization problem. There are numerous approaches. To select the most appropriate approach, it is essential to classify the modeled engineering optimization problem. Then, an algorithm that is suitable for each type of optimization problem can be selected. There are many different classification methods for optimization problems. Here, this book will introduce several of the most widely used in DfAM.

Discrete vs. Continuous: Some optimization models are designed to optimize the objective functions whose design variables take their allowed values from a "discrete set," which can be considered as a subset of integer values. For example, to optimize the ground truss structure shown in Figure 6.1a, each truss can be represented by a binary value. Specifically, a design variable vector $X = (x_1, x_2, \ldots, x_n)$ is defined. In this vector, x_i represents ith truss and only can be either 0 or 1. If $x_i = 0$, it means, this truss should be removed in the final optimized structure; otherwise, it should be kept. The overall compliance of structures can be considered as the function of these design variables. This is a typical discrete optimization problem since the design space is

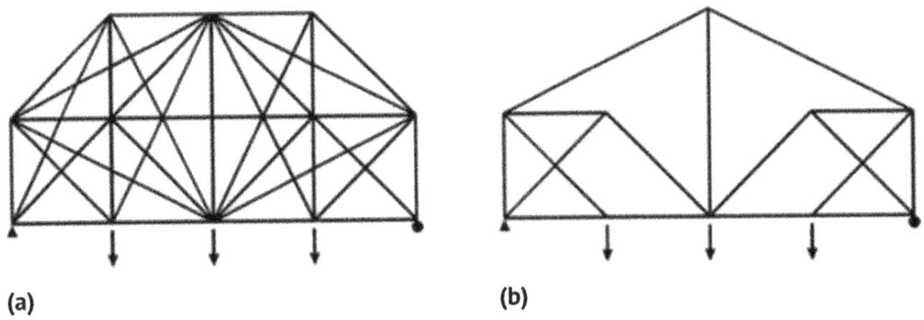

Figure 6.1 Topology optimization approach based on ground structure approach. (a) Ground structure and (b) final truss. *Source*: [1]/with permission of Elsevier.

not continuous. Compared to discrete optimization problems, the design variables of continuous optimization problems can take continuous real value. The example shown in Figure 6.1 can also be considered an example of the continuous optimization problem if we want to optimize the thickness or diameter of each strut in the truss. In this case, x_i defined in the design variables X refers to as the thickness of ith strut. Its values can continuously change from the lower bound to its upper bound. Discrete problems are often more difficult to solve due to their combinatorial nature and the lack of smoothness that continuous optimization can exploit, especially when the dimension of design variables is large. However, it should be noted that many discrete optimizations can be converted to a series of continuous subproblems. For example, relaxation techniques can be applied. Specifically, as in the example shown in Figure 6.1, for each strut, instead of using a discrete design variable $x_i \in \{0,1\}$, you can define a continuous variable $x_i \in [0,1]$. This converts the problem to a continuous problem where a gradient-based approach can be applied. After solving the problem, you may need to apply a heuristic method to convert the continuous value into a discrete solution. In addition, you can also modify the objective function and add the penalty term for the design variable's value between 0 and 1. This is the idea where one of the most popular topology optimization approaches Solid Isotropic Material with Penalization (SIMP) comes from.

Constrained vs. Unconstrained: Another common way to classify optimization approaches is based on whether there are constraints on the design variables. If there are no constraints, the problem is considered an unconstrained problem; otherwise, it is a constrained problem. Many engineering optimization problems are constrained, as there are often limitations that designers or practitioners must consider during decision-making. These constraints may arise from physical laws, manufacturability, cost, time, and other factors. It is worth noting that a constrained problem can be converted to an unconstrained problem using several widely used techniques, such as the augmented Lagrangian method.

Deterministic vs. Stochastic: In a deterministic optimization problem, all parameters, variables, and constraints are known with certainty, with no randomness involved. However, some data—such as parameters, constraints, or even the objective function—may involve randomness or uncertainty. This uncertainty can arise from incomplete or noisy information about the system. Most design optimization problems are considered deterministic since boundary conditions and design constraints are typically known. However, if manufacturing uncertainties lead to discrepancies between the designed structure and the manufactured structure, a stochastic optimization problem needs to be formulated to account for this uncertainty.

Convex and Non-convex: An optimization problem is called convex if the objective function is a real-valued convex function, and its feasible region is convex; otherwise, it is called a non-convex problem. In a convex problem, any local optimum is also the global optimum. Therefore, gradient-based approaches can be applied to find the global optimum for convex problems. In contrast, for non-convex problems, gradient-based approaches cannot guarantee convergence to the global optimum. As a result, heuristic approaches are often needed to find the global optimum in non-convex problems.

6.1.3 Algorithms for Engineering Optimization

Once the established optimization has been classified, algorithms that are suitable for this type of optimization problem can be selected and applied. There are numerous types of optimization algorithms. We are not going to discuss the mathematical foundations of those optimization algorithms. Instead, we will provide a classification and comparison of some of the most widely used algorithms for solving design optimization problems in DfAM. As to the detailed information for those optimization algorithms, readers can further reference the following textbooks [2, 3].

In general, optimization algorithms can be broadly classified into two major types: gradient-based and heuristic algorithms. This classification highlights the key differences in how these algorithms search for optimal solutions.

In gradient-based algorithms, the gradient of the objective function or Lagrangian function (for constrained problems) is used to guide the search for an optimal solution. This approach is effective for continuous, smooth, and differentiable problems. Typical gradient-based algorithms include gradient descent, conjugate gradient method, Newton's method, and quasi-Newton method such as BFGS (Broyden–Fletcher–Goldfarb–Shanno). Compared to heuristic algorithms, the gradient-based approach is efficient and generally faster for problems with well-behaved, differentiable objective, and constraint functions. For example, the Method of Moving Asymptotes (MMA) [4] is a widely used gradient-based optimization algorithm that has been used to solve the SIMP-based topology optimization problem. In the MMA approach, first-order derivatives of the objective and constraint functions are used to guide the building of the local approximation function that can be used to update the design variables. Its core idea is to approximate the original nonlinear optimization problem by a series of simpler subproblems that are convex and formed by introducing asymptotes that define bounds or limitations on how much the design variables can change in each iteration. The asymptotes control the step size for each variable, dynamically adjusting based on the progress of the optimization, which ensures stability. Compared to other gradient-based approaches, this approach is suitable for structural optimization, which usually has many design variables. It should be noted that all gradient-based optimization has limitations. For the non-convex problem, gradient-based optimization can only converge to the local optimal point. To improve their ability to meet the global optimum in non-convex problems, several modifications and techniques can be used. For example, the multi-start [5] approach can be used, which can increase the likelihood of finding the global optimum. However, it should be noted that, like other gradient-based optimization methods, these approaches can also not guarantee convergence to the global optimum.

Compared to gradient-based approaches, heuristic optimization approaches are more suitable for large nonlinear optimization problems that are not smooth or even continuous, where finding an exact solution is computationally expensive or infeasible. They leverage rules, intuition, or even trial-and-error techniques rather than strict mathematical procedures. Several typical heuristically optimization algorithms have been widely used for engineering optimization, such as genetic algorithms (GA) [6], simulated annealing (SA) [7], and particle swarm optimization [8]. These approaches use different strategies to update their design variables. These strategies are usually inspired by a natural or physical phenomenon. In those approaches, the parameters such as mutation rate in the GA algorithm play an important role balancing exploration and exploitation. Deciding on those parameters usually needs several rounds of trial and error.

6.2 Structural and Topology Optimization for AM

Among the various types of optimization methods, structural optimization methods are among the most widely used in design optimization. These methods help designers determine the size, shape, and topology of an optimal layout that achieves the best performance under specified boundary conditions. This section introduces the fundamentals of structural optimization techniques.

6.2.1 Introduction to Structural Optimization and Its Classification

A structural optimization algorithm is a type of computational optimization approach aimed at improving a structure's performance, efficiency, and material usage while satisfying design constraints. The goals of structural optimization algorithms can include reducing weight, cost, or material usage, or enhancing the structure's strength, durability, or other functional performance, such as thermal or electrical conductivity. Thanks to the unique design freedom offered by AM, structural optimization algorithms can be used to explore the expanded design space enabled by AM, further enhancing the performance of the designed part while maintaining or reducing its weight.

Generally, structural optimization can be divided into three major types: size optimization, shape optimization, and topology optimization. Size optimization focuses primarily on optimizing key dimensions of structural elements, such as thickness, length, and cross-sectional area of trusses, beams, or ribs. In this process, the algorithm adjusts these dimensions to achieve optimal performance based on specific criteria, such as minimizing weight or maximizing stiffness. As shown in Figure 6.2a, only the thickness of the truss is updated during the optimization process.

Compared to size optimization, shape optimization offers greater design freedom by allowing the algorithm to control the structure's shape. In Figure 6.2b, the shapes of holes are updated to provide better support under external loading. The final category is topology optimization, illustrated in Figure 6.2c. In this case, the initial design is a solid block without any internal holes. The topology optimization algorithm not only updates the shape of existing holes but can also create new ones, effectively altering the topology of the designed structure. Topology optimization can be seen as a combination of shape and size optimization, capable of generating more complex shapes that achieve better performance than the other two approaches.

However, topology optimization has its limitations. Without constraints on the shape or topology of the design space, it often generates geometries with complex features that are challenging to fabricate using conventional manufacturing methods such as CNC machining, casting, and injection molding. This limitation has been significantly mitigated by AM technologies, which allow the fabrication of complex geometries without additional cost. However, as discussed in Chapter 5, AM processes also have limitations, such as minimum feature sizes and support structure removal requirements. Compared to size and shape optimization, topology optimization offers less local control over feature details, making it more challenging to produce a structure that can be directly fabricated.

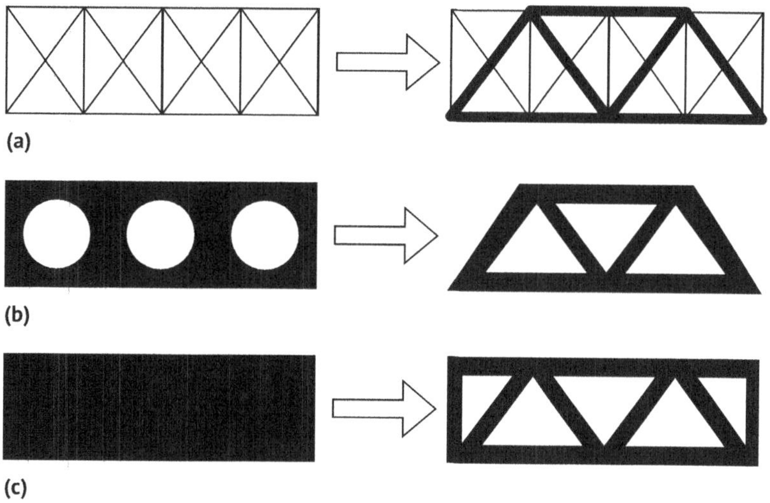

Figure 6.2 Comparison between different structural optimization methods: (a) size optimization, (b) shape optimization, and (c) topology optimization. *Source*: [9]/Springer Nature.

Significant research has been conducted to incorporate manufacturing constraints within topology optimization formulations, but a universal solution remains elusive, as constraints vary across different AM processes and materials. Therefore, solving a real design optimization problem often involves combining topology optimization with other optimization techniques, which will be discussed in detail in Section 6.2.2.

6.2.2 Topology Optimization

Due to its unique capabilities, topology optimization has garnered significant interest from designers seeking to enhance the functional performance of their designs. Various topology optimization methods exist, all of which modify structural topology to achieve improved performance. The primary distinction among these methods lies in how they parameterize the design space. Generally, based on the design space parameterization approach, topology optimization can be classified into density-based and non-density-based methods.

The density-based approach is the most widely used method in topology optimization. It represents material distribution within a continuum domain using a density field. Specifically, each element in the design space is assigned a value between 0 and 1, indicating the presence or absence of material. The most used density-based approach is known as SIMP. In SIMP, material stiffness is penalized for intermediate density values according to the following equation:

$$E(\rho) = E_0 \cdot \rho^p, \rho \in [0, 1] \tag{6.3}$$

where $E(\rho)$ is the effective elastic modulus at the element whose relative density is ρ. E_0 is the modulus of the solid material, and p is a penalization factor (typically $p > 1$) used to discourage the intermediate density and force design converging to either solid or void solution. Based on this parameterization scheme, the following optimization formulation can be formalized:

$$\begin{aligned}
\text{Min}: \quad & C(\rho) = \boldsymbol{F}^T\boldsymbol{U} = \textstyle\sum_{e=1}^{n} \rho_e^p u_e^T K_e u_e, \rho = (\rho_1, \rho_2, \ldots, \rho_e, \ldots \rho_n) \\
\text{S.T}: \quad & 0 < \rho_{\min} \leq \rho_e \leq 1, \forall e \in [1, n] \\
& \mathbf{KU} = \mathbf{F} \\
& \textstyle\sum_{e=1}^{e=n} \rho_e V_e \leq V_{\text{total}} = V_f \cdot \sum_{e=1}^{e=n} V_e
\end{aligned} \tag{6.4}$$

Where $C(\rho)$ is the objective function that can be used to calculate the compliance of a designed structure. Specifically, in the defined formulation, the finite element Method is used where the compliance can be calculated based on nodal force vector F and nodal displacement vector U. The design variable is the relative density vector $\boldsymbol{\rho}$, which consists of relative density ρ_e for each element of the design domain. In general, the relative density can be taken at any value from 0 to 1. However, to avoid the singularity of the stiffness matrix, usually during the implementation of the algorithm, a small value $\rho_{\min}$ which approach to zero is given as the lower bound. K is the structural stiffness matrix while K_e is the stiffness matrix for each element. The defined problem is constrained by the entire volume of the structure V_{total}. Sometimes, this constraint can also be described with the volume fraction V_f. To solve the defined optimization problem shown in Equation 6.4, the MMA algorithm described in Section 6.1.3 can be applied. Particularly, the gradient of the objective function can be calculated based on the following equation:

$$\frac{\partial C}{\partial \rho_e} = -p\rho^{p-1} u_e^T K_e u_e \tag{6.5}$$

The general step of solving SIMP-based topology optimization problem can be summarized as:

Step 1: Solve the FEA problem to obtain the displacement vector U
Step 2: Compute the strain energy $E_e = u_e^T K_e u_e$ for each element
Step 3: Calculate the compliance gradient based on Equation 6.5
Step 4: Combine the compliance gradient with the volume constraints gradient and update the design variables based on the MMA approach
Step 5: Check convergence; if it is not converged, go to Step 1 again.

For readers who are interested in implementing the SIMP method, this reference [10] is suggested to provide a more comprehensive guide on the implementation of the SIMP algorithm.

During the optimization process, filtering techniques can be applied to enhance the quality of optimized designs. Specifically, filtering helps avoid numerical instabilities and physically unrealistic designs, such as checkerboard patterns and small, unmanufacturable features. The filtering techniques used in topology optimization are generally divided into two types: density filtering and sensitivity filtering.

Density filtering is one of the simplest and most widely used methods. It directly applies to the design variables by averaging the densities of neighboring elements. Typically, this method is used after Step 4, where the smoothed densities are used to evaluate objective functions and constraints. Density filtering effectively prevents checkerboard patterns and produces a smoother design; however, it also has limitations, such as the loss of sharp boundaries and the inability to generate fine features. To address the issue of blurred boundaries, a Heaviside projection function can be used as a filter [11]. Compared to averaging filters, Heaviside projection allows designers to achieve sharper boundaries with fewer gray areas.

In addition to filtering relative density, designers can also filter sensitivities. Sensitivity filtering smooths the gradients of the objective function and helps prevent local irregularities in the sensitivity field, which can cause numerical issues or unrealistic designs. Unlike density filtering, sensitivity filtering provides less control over the design variables but helps reduce large oscillations during the optimization process.

Besides the SIMP method introduced above, there are also many other topology optimization methods, such as the level-set method [12, 13], bidirectional evolution structural optimization (BESO) [14], etc. In this book, we are not going to discuss each type of topology optimization method. Readers who are interested in them can read the relevant references cited in this book. Among all the different types of methods, SIMP has been most widely used in commercial software due to its easy implementation based on the existing finite element simulation framework. However, it also has several limitations, such as mesh dependence and the gray region usually causes unsmooth boundaries. In addition, for all topology optimization methods, it is usually difficult to directly consider the manufacturing constraints in the design process. To solve this challenge, a lot of research has been done recently using different approaches to integrate the design constraints described in Chapter 5 in the topology design of AM-manufactured structures. Even though the manufacturability of optimized structure can be improved by those approaches, those additional constraints may also lead the optimization problem to converge to the local optimal, which is the issue of all gradient-based approaches during the solving a non-convex optimization problem. In addition, designers also need to double-check the optimized output and ensure it is manufacturable.

6.2.3 Apply Topology Optimization in DfAM

Currently, topology optimization tools have already been implemented in most CAD/CAE software. However, it is still imperative for designers to understand how to correctly use this tool to support the design, since the same tool may generate different design results if the designers fail to identify the proper design freedom and constraints. It is generally suggested to follow the steps listed below:

Step 1: Define the design space and non-design space
Step 2: Define the topology optimization problem
Step 3: Run TO tool to generate the optimal design layout
Step 4: Smooth the optimal results and build the CAD model based on the optimization result
Step 5: Run either size or shape optimization to further optimize the structure
Step 6: Final FEA check of the performance of the optimized structure

Among the suggested steps discussed above, defining the design and non-design space plays a critical role. As shown in Figure 6.3, for the same design, different sizes of design space are defined for different scenarios, which leads to different optimal topology layouts as well as functional performance. In this book, a functional-driven approach is suggested to help designers define the design space. Specifically, functional analysis is suggested at the beginning of the design. Based on functional analysis, functional surfaces (FSs) and functional volumes (FVs) need to be defined. Here, FS represents the key surface for certain functional requirements during the functional analysis, while FV refers to the solid region that connects the FSs and assists FSs in achieving the given function. An example of a functional analysis result is shown in Figure 6.4. 5 FSs play roles as connection and assembly functions for this structural component. To connect those FSs, one FV is designed. The FSs need to be kept in the optimized design; therefore, the non-design space can be obtained by thickening the FSs. The rest of FV is a non-design space.

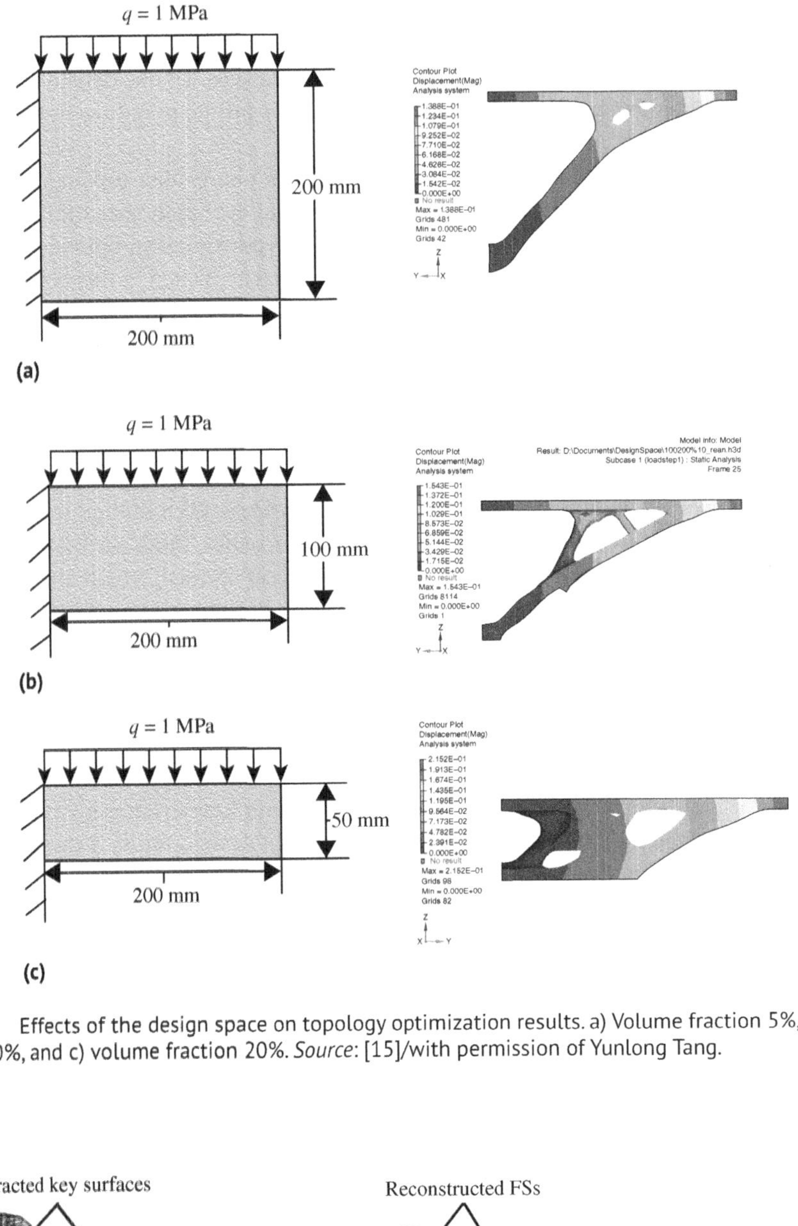

Figure 6.3 Effects of the design space on topology optimization results. a) Volume fraction 5%, b) volume fraction 10%, and c) volume fraction 20%. *Source*: [15]/with permission of Yunlong Tang.

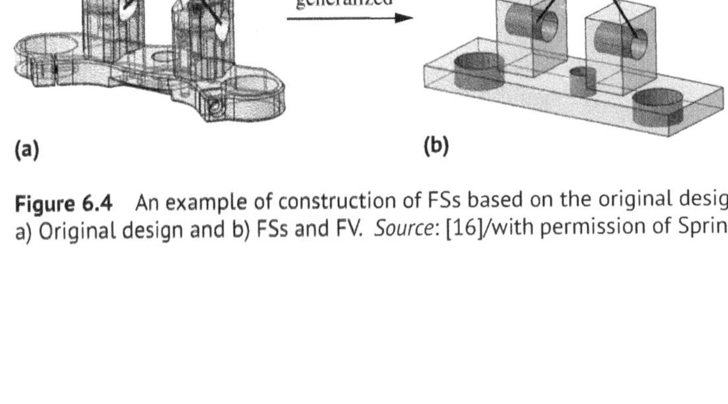

Figure 6.4 An example of construction of FSs based on the original design. a) Original design and b) FSs and FV. *Source*: [16]/with permission of Springer Nature.

In the second step, the topology optimization problem needs to be defined. Typically, compliance is set as the objective with a given weight or volume fraction constraints. However, it should also be noted that the structural weight itself can also be set as the objective function. In this case, compliance can be set as a constraint. There is typically a trade-off between minimizing weight and ensuring structural stiffness (low compliance). Designers may run several topology optimizations with different constraint values to find the optimal solution that can meet all the requirements. Besides the volumetric constraints, other constraints such as manufacturability or even stress constraints can be added to the topology optimization formulation. It should be noted that even those constraints are added but cannot fully guarantee the final design can meet stress and manufacturability requirements because of gray density region. Thus, designers still need to run a full FEA simulation to check the stress and use the design guideline to check the manufacturability constraints.

The output of topology optimization results is usually represented by a coarse triangular mesh. Thus, after running the topology optimization, it is always necessary to smooth the boundary of topology optimization results. Many TO tools or software provide this function. However, the smooth mesh may also be difficult to modify locally. To solve this issue, the (boundary representation) BREP model needs to be built based on optimized results. The rebuilt BREP model makes it much easier to control its geometry locally, which offers freedom for further design and optimization. Also, during the rebuild of geometry, the manufacturability constraints can be considered to further improve the manufacturability of designed parts. To illustrate this step, an example of mesh directly output from topology optimization is shown in Figure 6.5a. In this part, the internal cavity is not visible due to the thin layer of skin material on the top. Based on the material distribution, the rebuilt BREP model is shown in Figure 6.5b. This modified model can be further optimized by parametrically controlling its shape.

Based on the rebuild BREP component, another round of size and shape optimization can be applied as stated in Step 5. Finally, the FEA tool will be used to validate the final performance of the optimized component.

6.2.4 Multi-scale Structural Design Based on Topology Optimization

Structural design optimization cannot only be applied on the macroscopic, it can also be applied on the meso or micro scale. For example, the inverse homogenization technique [18] can be used to optimize the material layout in the unit cell to achieve particularly given effective material

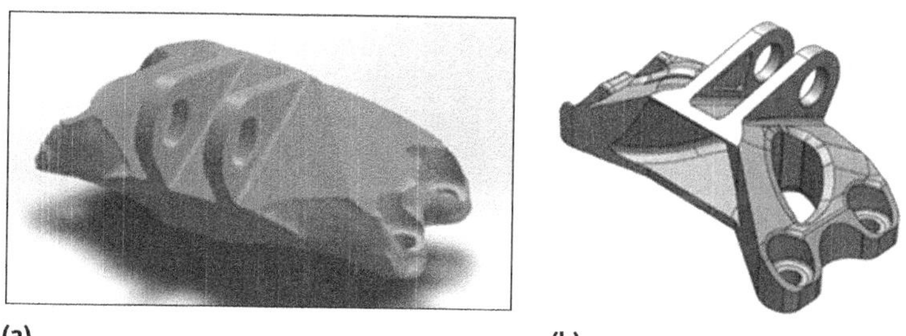

(a) (b)

Figure 6.5 Rebuild the CAD model for topology optimization. a) CAD after topology optimization and b) rebuild CAD model based on topology optimization. *Source*: [17]/with permission of Elsevier.

properties. In this problem, the homogenization algorithm can be used to calculate the effective properties of the unit cell, while the topology optimization method is used to update the topology of the unit cell. This approach has been used to design both 2D and 3D porous structures. Besides the design of structures on different scales independently, the concurrent topology optimization methods [19] also enable the design and optimize structures on different scales simultaneously. For example, the recent topology optimization approach enables the optimization of the fiber layout for additively manufactured continuous fiber-reinforced composites as shown in Figure 6.6 [20]. Its results show a significant increase in structural performance by fully utilizing the design freedom enabled by AM from different design scales.

6.3 Multi-disciplinary Design Optimization for AM

In addition to optimizing structural performance, the unique design freedom provided by the AM process presents a valuable opportunity for designers to enhance multiple functions simultaneously. For example, the porous lattice structures can be optimized for both heat conduction and load-bearing functions [15]. However, the design parameters for lattice structures may affect the performance of both functions at the same time. In some cases, the performance of one function might also determine the performance of other functions. Those functions cannot be fully decoupled. To solve these issues, this book introduces an approach called F-P-P-D (Function–Performance–Property–Design parameter) model. Its general graphical representation is shown in Figure 6.7.

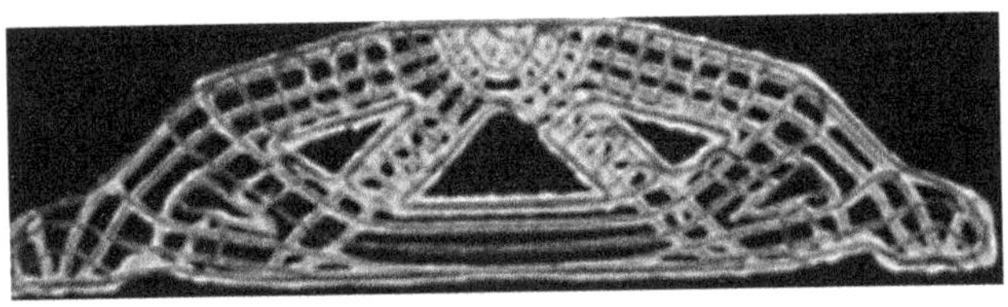

Figure 6.6 Example of continuous fiber-reinforced structure with multi-scale topology optimization. *Source*: [20]/with permission of Elsevier.

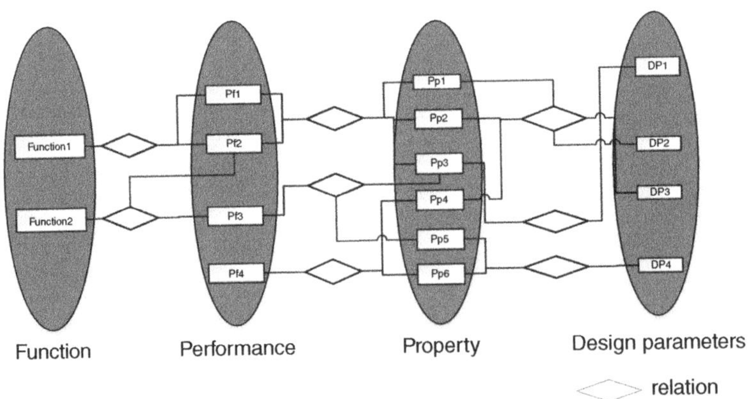

Figure 6.7 Graphic view of F-P-P-D relationship. *Source*: [15]/with permission of Yunlong Tang.

As illustrated in Figure 6.7, there are four major domains in F-P-P-D model. The first domain is called as the functional domain. It saves the names of major functions that the designed structures plays, such as load bearing, heat dissipation or insulation, energy absorption, damping, or even the electromagnetic wave absorption.

To quantitatively describe the function defined in the functional domain, the performance domain is defined in the F-P-P-D model. This domain specifies the detailed requirements for the performance of each function. For example, for a structural load-bearing function, designers can define the maximum displacement and allowable stress as performance parameters. These parameters can be evaluated and optimized during the design process and directly relate to the success criteria for their respective functions. In addition to these performance parameters, other factors such as weight and cost are also important to consider during the design of structures fabricated by AM. These parameters should also be listed in the performance domain.

To achieve the desired performance parameters, the related properties of a designed structure can be controlled. Thus, we proposed a third domain called the properties domain. The effective properties of designed lattice structures can be defined in this domain, such as the effective elastic modulus and heat conduction coefficient. By varying the relative density and even the topology of meso-scale lattice structures, these properties can be nonuniformly distributed within the given geometric boundary. A field function can then be defined to describe their distribution.

The final domain of the proposed F-P-P-D model is the design parameters domain. The design parameters of lattice structures, including relative density, unit cell size, lattice orientations, and even lattice topology can be defined in this domain to control the properties in the properties domain. These parameters can also be varied in three-dimensional space and can be described by a field function defined in the three-dimensional design space.

Beyond the entities defined within each domain, the relationships among these entities are even more critical. For example, each function is linked to its corresponding performance parameters, as illustrated in Figure 6.7. Each performance parameter is, in turn, connected to the properties that influence its value. Similarly, each property is associated with the design parameters that determine its behavior.

In multifunctional design, functions may become coupled at the performance, property, or even design-parameter levels. Based on the F-P-P-D model introduced above, a multidisciplinary design optimization (MDO) formulation can be constructed. This allows the optimal distribution of design parameters to be obtained for all functions defined in the function domain. Readers interested in the detailed process of converting the F-P-P-D model into an MDO formulation may refer to Reference [15]. After solving the defined MDO formulation, the optimized distribution of lattice parameters is obtained, enabling the design to achieve optimal performance for both functions. An example design case is shown in Figure 6.8. In this case, the panel serves two roles. First, conducting heat from the center region of the panel (highlighted by the red line) to warm the gas beneath it, while preventing heat from being transferred to the two sidewalls that support the panel. At the same time, the panel must be capable of withstanding the high gas pressure applied from below. Based on the established F-P-P-D model, the corresponding MDO formulation is solved, resulting in an optimized spatial distribution of lattice relative density. As illustrated in Figure 6.8, the heterogeneous lattice structure significantly enhances both structural and thermal performance—reducing compliance while improving heat-conduction efficiency.

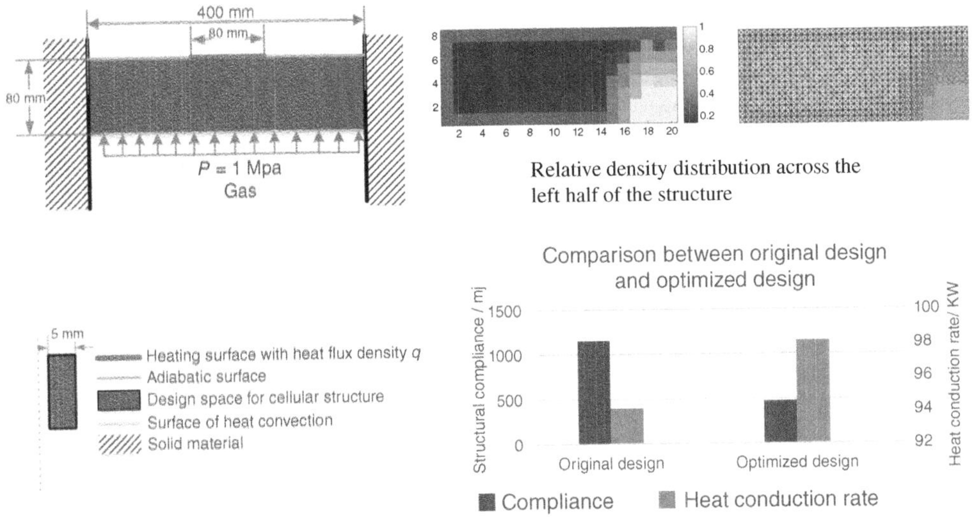

Figure 6.8 A porous lattice panel designed for both heat conduction and load bearing. *Source*: [15] / with permission of Yunlong Tang.

6.4 Reliability-based Design Optimization

AM has transformed the production of complex, high-performance parts by offering unprecedented design freedom. However, AM processes also face certain limitations, particularly in terms of process repeatability and stability. For load-critical applications, such as load-bearing parts for aerospace and biomedical fields, ensuring the reliability and quality of AM-fabricated components is essential. In these industries, failures can have severe consequences, leading to costly maintenance, system downtime, or even safety hazards.

RBDO directly addresses the challenges associated with uncertainty. It aims to optimize structural performance while accounting for various sources of uncertainty, including material properties, geometric deviations, load conditions, and environmental variations. This approach ensures that the final design is not only optimized for performance but also robust enough to meet reliability targets under real-world conditions. For instance, in the design of an aerospace component fabricated through AM, ensuring the reliability of the component is critical to avoid catastrophic failures. During fabrication, uncontrolled variations in the AM process—such as fluctuations in laser power, scanning speed, and powder bed density—may affect material quality. Additionally, loading conditions may change during the service life of the component. To account for these uncertainties, the RBDO approach can be applied, incorporating failure probability into the optimization formulation. This probability can be calculated based on the following equation:

$$P_f = P_r[G(x,y) \leq 0] = \int_{G(x,y \leq 0)} f_Y(y) dy_1 \dots dy_n \tag{6.6}$$

where P_f is the failure probability, $f_Y(y)$ is the joint function of the random variables Y. Here the random variables can include the loading directions, the variation of structural geometry or material properties induced by the AM process. $G(x,y)$ is defined as a limit state function. Specifically, $G(x,y) < 0$ represents the failure state and the state of safety $G(x,y) > 0$. In structural optimization, the failure condition can be expressed based on the limitation of maximum displacement or maximum stress that materials can handle without plastic deformation. The evaluation of P_r based on Equation 6.6 is difficult due to not all the information of the joint probability density function is available. Thus, the first- and second-order reliability methods (FORM/SORM) [21] were developed based on the reliability index concept [21]. A generalized probability index β_G can be defined as:

$$P_r(G) = \Phi(-\beta_G) \tag{6.7}$$

where $\Phi(\cdot)$ is the standard normal cumulative distribution function. The general form of the RBDO problem can be expressed as:

$$\begin{aligned} \text{min:} \quad & \text{f(x)} \\ \text{s.t.:} \quad & P_{fj} = P_r\left(G_j(x,y) < 0\right) \leq \overline{P_{fj}}, j = 1,2,\ldots,np \end{aligned} \tag{6.8}$$

where $\overline{P_{fj}}$ is the prescribed failure probability limit for jth constrain, y is a vector of random variables. The constraints in Equation 6.8 can be rewritten into the following form based on the reliability target index.

$$\beta_s = -\phi^{-1}(P_r(0)) \geq \beta_t \tag{6.9}$$

In this context, β_t represents the target reliability index. The RBDO problem can be incorporated into the topology optimization problem described in Section 6.3, where variations in load conditions and geometries are considered. Generally, the RBDO problem can be viewed as a specialized type of multi-objective optimization problem, as it involves balancing both structural performance and reliability index. These two aspects are also interdependent: design variables influence not only structural performance but also impact the reliability index. Random variables are critical for the reliability index but do not play the same role in the problem's objective function.

As a result, solving the RBDO problem typically requires a higher computational load compared to deterministic structural optimization. For readers interested in further exploring methods to solve RBDO problems, additional references are provided [21, 22].

6.5 Conclusion

Engineering optimization, particularly structural optimization methods, is an effective tool for helping designers explore the expanded design freedom enabled by the AM process. In this chapter, the fundamentals of engineering structural optimization and its application in DfAM have been introduced. Designers can use this tool not only to improve the structural performance of designed components but also to simultaneously enhance performance across different functions.

During optimization, it is important to consider manufacturing constraints and uncertainties from the AM process. Specifically, manufacturability can be addressed at various stages of structural optimization. As for uncertainties, RBDO is an effective tool for designing structures that play critical roles.

References

1 Deb, K. and Gulati, S., *Design of truss-structures for minimum weight using genetic algorithms*. Finite Elements in Analysis and Design, 2001. **37**(5): p. 447–465.

2 Antoniou, A. and Lu, W.-S., *Practical optimization: algorithms and engineering applications*. Vol. 19. 2007, Springer.

3 Rao, S.S., *Engineering optimization: theory and practice*. 2019, John Wiley & Sons.
4 Svanberg, K., *The method of moving asymptotes—a new method for structural optimization*. International Journal for Numerical Methods in Engineering, 1987. **24**(2): p. 359–373.
5 Jain, P. and Agogino, A., *Global optimization using the multistart method*. Journal of Mechanical Design, 1993. **115**(4): pp. 770–775.
6 Lambora, A., Gupta, K., and Chopra, K., Genetic algorithm-A literature review. in *2019 International Conference on Machine Learning, Big Data, Cloud and Parallel Computing (COMITCon)*. 2019. IEEE.
7 Bertsimas, D. and Tsitsiklis, J., *Simulated annealing*. Statistical Science, 1993. **8**(1): p. 10–15.
8 Kennedy, J. and Eberhart, R., *Particle swarm optimization*. in *Proceedings of ICNN'95-International Conference on Neural Networks*. 1995. IEEE.
9 Bendsoe, M.P. and Sigmund, O., *Topology optimization: theory, methods, and applications*. 2013, Springer Science & Business Media.
10 Sigmund, O., *A 99 line topology optimization code written in Matlab*. Structural and Multidisciplinary Optimization, 2001. **21**(2): p. 120–127.
11 Kawamoto, A., et al., *Heaviside projection based topology optimization by a PDE-filtered scalar function*. Structural and Multidisciplinary Optimization, 2011. **44**: p. 19–24.
12 Wang, M.Y., Wang, X., and Guo, D., *A level set method for structural topology optimization*. Computer Methods in Applied Mechanics and Engineering, 2003. **192**(1): p. 227–246.
13 Van Dijk, N.P., et al., *Level-set methods for structural topology optimization: a review*. Structural and Multidisciplinary Optimization, 2013. **48**: p. 437–472.
14 Huang, X. and Xie, Y.M., *Bi-directional evolutionary topology optimization of continuum structures with one or multiple materials*. Computational Mechanics, 2009. **43**: p. 393–401.
15 Tang, Y., *A multiscale and multifunctional design method for mesoscale lattice structures fabricated by additive manufacturing processes*. 2017, McGill University (Canada).
16 Tang, Y., Yang, S., and Zhao, Y.F., *Sustainable design for additive manufacturing through functionality integration and part consolidation*. in *handbook of sustainability in additive manufacturing: Volume 1*, S.S. Muthu and M.M. Savalani, Editors. 2016, Springer Singapore: Singapore. p. 101–144.
17 Tang, Y., Mak, K., and Zhao, Y.F., *A framework to reduce product environmental impact through design optimization for additive manufacturing*. Journal of Cleaner Production, 2016. **137**: p. 1560–1572.
18 Sigmund, O., *Materials with prescribed constitutive parameters: an inverse homogenization problem*. International Journal of Solids and Structures, 1994. **31**(17): p. 2313–2329.
19 Liu, P., Kang, Z., and Luo, Y., *Two-scale concurrent topology optimization of lattice structures with connectable microstructures*. Additive Manufacturing, 2020. **36**: p. 101427.
20 Liu, T., et al., *Stress-driven infill mapping for 3D-printed continuous fiber composite with tunable infill density and morphology*. Additive Manufacturing, 2023. **62**: p. 103374.
21 Hasofer, A.M., *An exact and invarient first order reliability format*. Journal of the Engineering Mechanics Division, Proceedings of the American Society of Civil Engineers, 1974. **100**(1): p. 111–121.
22 Kharmanda, G., et al., *Reliability-based topology optimization*. Structural and Multidisciplinary Optimization, 2004. **26**(5): p. 295–307.

7

Lattice Structure Design and Simulation

In Chapter 6, we introduced topology optimization and structural optimization within the context of design for additive manufacturing (DfAM). Among the various structural forms employed in topology optimization for additive manufacturing (AM), **lattice structures** are particularly prominent. These structures have their origins in nature—wood, bone, and honeycomb are all examples of natural lattice structures (or cellular materials) found at different length scales. Inspired by these biological forms, engineers and researchers began exploring the use of lattice structures in mechanical and structural design nearly half a century ago, well before the emergence of AM. However, it is only in the past decade that lattice structures have gained significant attention, largely due to the capabilities of AM to fabricate complex geometries that were previously impractical or impossible to manufacture using conventional methods. Lattice structures have been widely adopted in function-driven design applications, including lightweighting, energy absorption, thermal management, and medical implants. This chapter begins by addressing the fundamental question: *What is a lattice structure, and how is it fabricated using AM?* Then, this chapter presents a general design framework and simulation-based design methodologies tailored for AM-produced lattice structures. Finally, the topology optimization techniques specifically developed for lattice structure design are discussed.

7.1 Introduction to Lattice Structures

The lattice structure in this book refers to a type of cellular architectures that has truss-like structures with interconnected struts and nodes in a 3D space [1]. Before the existence of AM, the geometry of lattice structures is quite simple due to manufacturing constraints of conventional manufacturing methods. AM has largely released these manufacturing constraints for lattice structures design. Compared to other cellular materials such as random foams or honeycombs, the lattice structure exhibits better mechanical performance [2]. By tailoring the material, the lattice structure cam be designed with a variety of physical properties including lightweight [3], nearly zero thermal expansion coefficient [4], negative Poisson's ratio [5], and high surface-to-volume ratio for heat dissipation via active cooling [6]. lattice structures have been extensively implemented in engineering applications to exploit the promising physical and mechanical properties, such as lightweight structures [7, 8], energy absorption [9], low thermal expansion structures [10], and conformal cooling [11]. Furthermore, lattice structures have potential if the application of biocompatible materials that need customized porosity and stiffness for orthopedic implant [12] and tissue engineering [13–15].

7.1.1 Lattice Structure Classification

According to literatures [7, 16, 17], lattice structures can be classified using various criteria. In this chapter, they are categorized based on the degree of order in the lattice framework, which generally yields three types. The first type, referred to as disordered or randomized lattice structures, consists of unit cells that are distributed randomly throughout the design space, with variations in both topology and cell size. An example of this type is shown in Figure 7.1a. The second type is the periodic lattice structure, illustrated in Figure 7.1b. These structures are formed by the regular, repeating arrangement of identical unit cells in three-dimensional space, such that each cell has the same topology and size. The third type is the pseudo-periodic lattice structure, in which all lattice cells share the same topology but differ in size and shape, providing a controlled variation between fully periodic and fully randomized arrangements. For example, the conformal lattice structure, which was first proposed by Wang and Rosen [18], is a typical pseudo-periodic lattice structure. Conformal lattice structures can keep the integrity of their unit cells on the boundary, as shown in Figure 7.1c, which is able to stiffen or strengthen a complex and curved surface better than periodic lattice structures [19, 20].

Both the periodic and the pseudo-periodic lattice structures can be further divided into two subtypes based on the uniformity of strut's thickness. They are homogeneous lattice structures and heterogeneous lattice structures. If the strut thickness is properly optimized, the heterogeneous lattice structure will exhibit better mechanical performance than the homogeneous lattice structure [21]. For example, in Figure 7.2, two drone arms are infilled with homogeneous and heterogeneous lattice structures. The heterogeneous one is optimized to increase the stiffness of the structure while maintaining the weight [22].

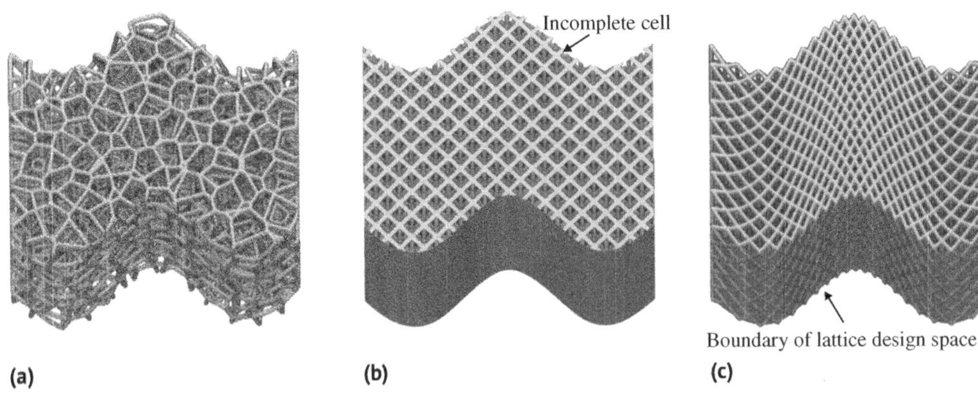

Figure 7.1 Examples of different types of lattice structures based on the degree of order. (a) Disordered lattice structures, (b) periodic lattice structures, and (c) conformal lattice structures.

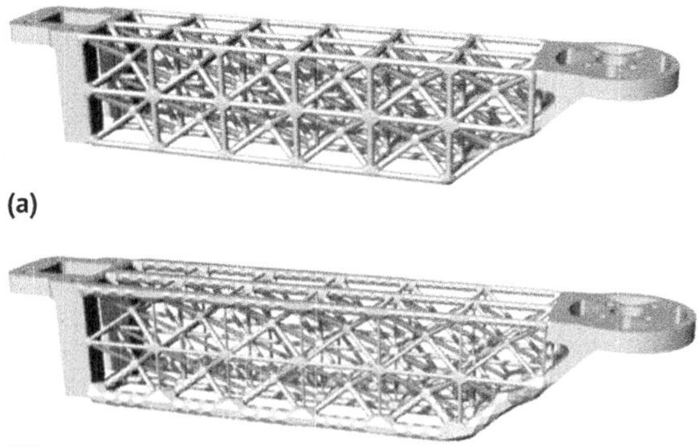

(a)

(b)

Figure 7.2 Examples of homogeneous and heterogeneous lattice structures. (a) Homogeneous lattice structure and (b) heterogeneous lattice structure.

7.1.2 Manufacturing Methods for Lattice Structures

Lattice structures with complex geometries can be successfully fabricated by several manufacturing techniques. Among them, AM is especially well suited for the fabrication of complex lattice structures due to its layer by layer fabrication strategy. Lattice structures designed for engineering applications have been successfully fabricated by different types of AM techniques such as stereolithography [23], fused deposition modeling [24, 25], binder jetting [26], selective laser melting [27], selective laser sintering [28], and electron beam melting [29, 30]. Schaedler and Carter [31] reviewed and compared various AM techniques for lattice structures, considering factors such as geometry, material, graded properties, feature size, application performance, and production rate. AM can also be combined with metal casting, a process known as "indirect AM-based casting," to produce lattice structures [32]. Although this approach involves additional fabrication steps, it alleviates several limitations of conventional AM, including thermal residual stresses, poor surface finish, and restricted material selection. Recently, ultralight metallic micro-lattices have been fabricated by an indirect AM method. It deposited thin metallic films onto polymer templates created via AM, which are then removed through chemical etching [33]. Apart from AM, lattice structures can also be fabricated by traditional manufacturing methods. One type of 3D carbon fiber reinforced plastic lattice is assembled by two-dimensional building blocks of [0/90] carbon fiber reinforced plastic laminate with reversible snap-fit connectors [34]. Three-dimensional metallic and composite lattices can also be fabricated by weaving, braiding, and knitting fibers and wires. However, these approaches are only able to manufacture lattice structures with simple topology. AM is still the most ideal and prevalent way to manufacture lattice structures currently.

It is found that the manufacturing characteristics of AM processes have considerable influence on the mechanical properties of manufactured lattice structures. Firstly, the irregularity of the surface finishing is one of the most common geometrical discrepancies of the AM-fabricated lattice structures. This phenomenon is caused by both the stair-step effect and unmelted powders on the surface of parts, as shown in Figure 7.3. Secondly, the anisotropic material property of AM-fabricated parts is another important aspect that needs to be considered in design and simulation of lattice structures. It is found in the literature that the mechanical properties are different in different orientations among most AM processes. To properly model the manufacturing influence in the simulation is critical. This will guarantee more accurate simulation results.

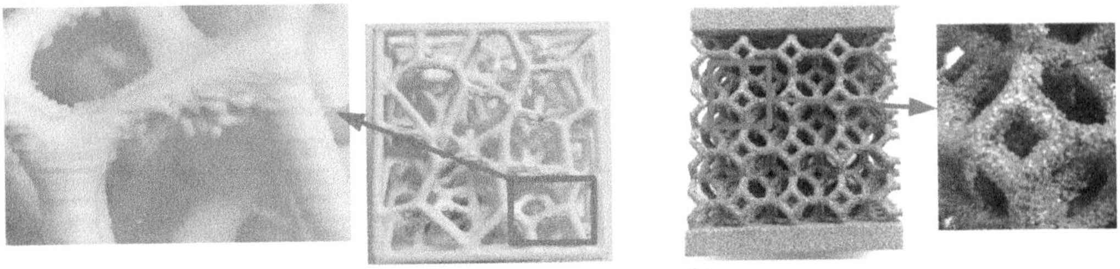

(a) (b)

Figure 7.3 Surface irregularity caused by stair-step effects and unmelted powder. (a) Stair-step effects and (b) unmelted powder.

7.2 Design Methodology for Lattice Structures

There are three main design stages, initial design, lattice design, and design validation, for solid–lattice hybrid structures. As shown in Figure 7.4, functional analysis is conducted in the initial design stage to separate the original model into design and non-design spaces. Topology optimization is implemented to determine the importance (element density) of the material in the design space. The high-density region is defined as the solid space, and the low-density region is defined as the lattice space. Next, the solid space is merged with the non-design space to generate the solid part of the final design. In the lattice design stage, lattice structures are generated within the lattice space by selecting the topology and unit cell size and by the modification of the lattice wireframe as needed. The optimization in this stage updates the strut thickness in the lattice structure to improve its stiffness. After that, the lattice wireframes are thickened by the optimal strut diameters. The optimized lattice structure is then combined with the solid part as a solid–lattice hybrid structure. The whole structure is tested in the validation stage to evaluate the functional requirement satisfaction. A control arm of the automobile suspension is used as an example to discuss the details of the design flow.

7.2.1 Initial Design

The initial design stage consists of three steps. The first step involves determining the design space and non-design space through an evaluation of the functional surfaces (FSs) and functional volumes (FVs). FSs are defined as any surfaces that must be preserved because it provides a required function, such as load transfer, assembly alignment, motion constraints, or interface geometry. FVs include the material regions that connect to and structurally support these FSs. These volumes are necessary for the FSs to operate correctly and therefore must be retained during the optimization process [7]. In the control arm example, three surfaces serve as assembly interfaces with the surrounding suspension components. These surfaces are therefore identified as FSs and must be preserved throughout the design process. The material regions that connect to and support these FSs constitute the FVs, which play essential roles in ensuring proper assembly and load transfer within the suspension system. These FVs are designated as the non-design space, while all remaining portions of the original component are defined as the design space, as shown in Figure 7.5.

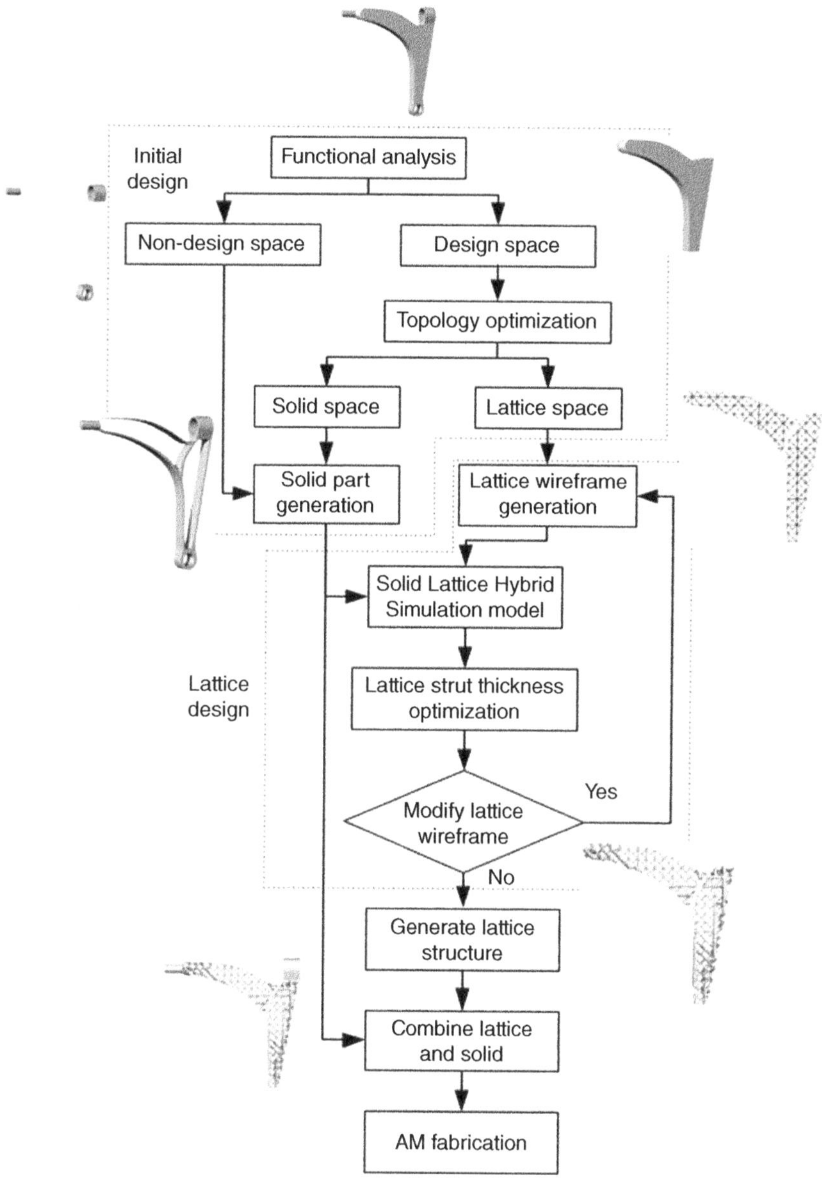

Figure 7.4 The design methodology for lattice structures. *Source*: [1]/with permission of Elsevier.

The second step of the initial design process involves performing topology optimization to determine a more efficient distribution of material. The initial control arm geometry contains large regions of material that contribute little to its structural performance and therefore represent opportunities for weight reduction. Topology optimization is a powerful computational technique that identifies the optimal placement of material within the design space based on loads, constraints, and performance objectives. The design space is discretized into finite elements, and each element is assigned a design variable, typically a density between 0 (void) and 1 (solid). An

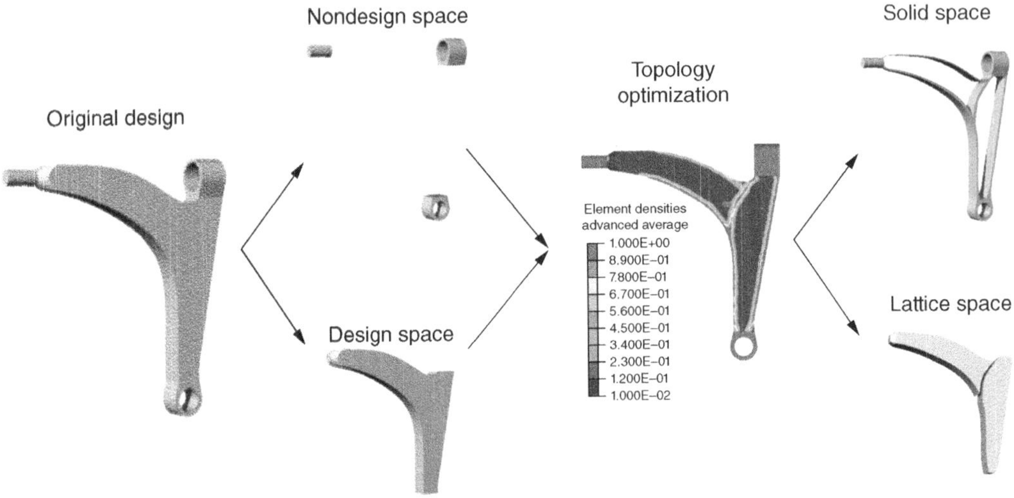

Figure 7.5 Initial design to define the solid space and lattice space. *Source*: [1]/with permission of Elsevier.

optimization algorithm then iteratively updates these densities to maximize structural stiffness or minimize compliance while satisfying constraints such as volume fraction. By concentrating material in load-bearing regions (increasing the density) and removing it from less critical areas (reducing the density), the method enhances the stiffness of the structure while preserving or reducing its mass. In this example, HyperWorks OptiStruct is employed to perform the topology optimization on the design space. The maximum allowable material usage is constrained to 20% of the original weight. As shown in Figure 7.5, the optimized design contains large regions of low relative density, indicating areas where the solid material can be effectively replaced with lattice structures.

The third step involves segmenting the solid and lattice spaces based on the topology optimization results from the previous step. The following principles guide the generation of the solid space:

1. **Retention of high-density regions**: Areas with high relative density are preserved as solid space.
2. **Density threshold adjustment**: The density threshold can be tuned to balance weight reduction and stiffness requirements. A higher threshold reduces the volume designated as solid space.
3. **Connection to functional elements**: FVs and FSs must remain linked to the solid space.
4. **Structural integrity**: The solid space must maintain continuity. If a high-density threshold fragments the solid region, the parts should be reconnected to ensure proper load transfer.

For instance, according to the density distribution shown in Figure 7.5, the solid space is redesigned in CAD software to connect the non-design space in the control arm. The remaining intermediate and low-density areas are set to lattice space, which can sustain the integrity and robustness of the control arm and support the solid part during the AM fabrication process. The lattice space defines the volume for the lattice wireframe generation.

7.2.2 Lattice Design

In the lattice design stage, wireframe of the lattice structure is generated at the beginning. The topology and unit cell size determine the geometry of the wireframe. A uniform lattice wireframe is generated in the bounding box of the lattice space as shown in Figure 7.6a–c and is then trimmed by the boundary surfaces of the lattice space as shown in Figure 7.6d. It should be noted, however, that trimming the lattice wireframe can introduce several issues. Some nodes may lie very close to the boundary of the lattice space. If a node falls outside the boundary, it is removed, and any struts connected to it are trimmed, as illustrated in Case 1 of Figure 7.6d. This can reduce the structural strength of the lattice because disconnected struts no longer contribute to load transfer. In the other case, as shown in Case 2 of Figure 7.6d, nodes that are inside the boundary but very close to it may leave short struts along the boundary. It can cause simulation difficulty because beam element is used to model the lattice structure in the optimization process. Typically, the aspect ratio of lattice struts ranges from 1:20 to 1:5, and the lattice is often modeled using Timoshenko beam elements. While the Timoshenko beam accounts for shear deformation, struts that are too short cannot be accurately represented by beam theory. Consequently, these short, trimmed struts may introduce errors into the simulation results if beam elements are used. In both cases A and B, the connecting points will be moved to the boundary to avoid the issues. The unconnected struts in Case 1 are connected, and the short struts in Case 2 are removed.

The objective of the lattice design stage is to adjust the relative density of lattice structures to satisfy the functional requirements as well as improving the efficiency of the structure (e.g., reduce the weight). Because the manufacturing is expensive and time-consuming, numerical simulation methods are usually used to predict the performance of lattice structures in the detailed design stage. The most popular methods to predict the mechanical performance of lattice structures are the homogenization method and finite element method (FEM). The details

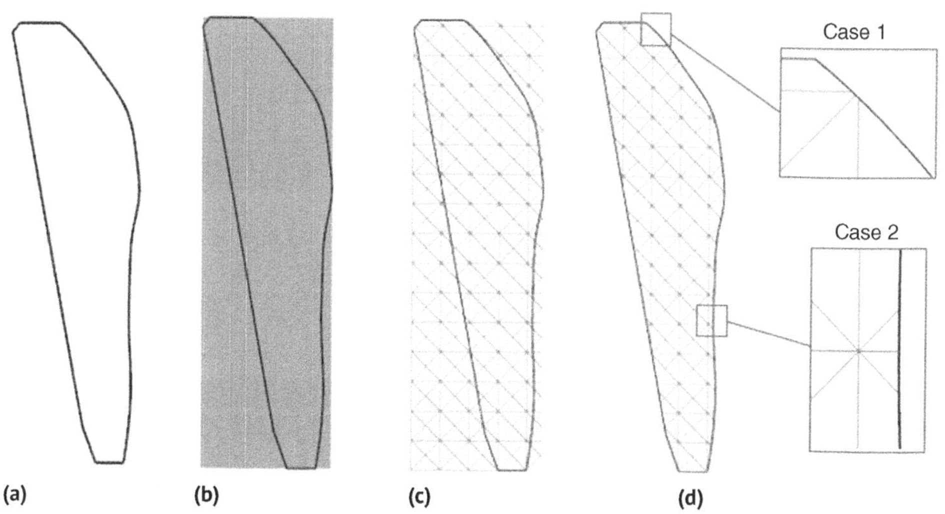

Figure 7.6 Generation of the lattice wireframe. (a) Lattice space, (b) bounding box, (c) wireframe, and (d) trimmed wireframe. *Source*: [1]/with permission of Elsevier.

about the simulation of lattice structures will be introduced in Section 7.3. In general, the design process is driven by the simulation models. The stress and strain predicted by the simulation model can be indicators to adjust the relative density or strut thickness of lattice structures. Optimization algorithms have been developed to find the optimal design of lattice structures under certain design requirements. In Section 7.4, two optimization methods will be introduced in detail. Once the optimal result is found, the information will be used to generate the final design of lattice structures.

7.2.3 Design Validation

The last stage in the process is to validate the design result. Although the simulation model can estimate the mechanical performance of lattice structures. It cannot guarantee the strength of the structure due to the defects existing in the as-built structures. Physical experiments will be conducted to test the real mechanical performance of the optimized structure. The result will be compared with the simulation result in the detailed design process. An example of the design validation is shown in Figure 7.7. In general, a specific fixture needs to be designed to add the boundary conditions to the optimized structure. The stiffness and strength of the structure can be obtained from the physical experiment. If there is a discrepancy between experimental result and the simulation result, both the simulation model and the optimization algorithm should be checked. The simulation model should be adjusted to reduce the error between estimated and real performance. The parameters such as the minimum and maximum strut thickness defined in the optimization algorithm need to be revised to satisfy the manufacturing constraints. After the simulation model and optimization algorithm are tuned, the detailed design process will generate the next round of design for the validation process until the structure passes the physical experiment examination.

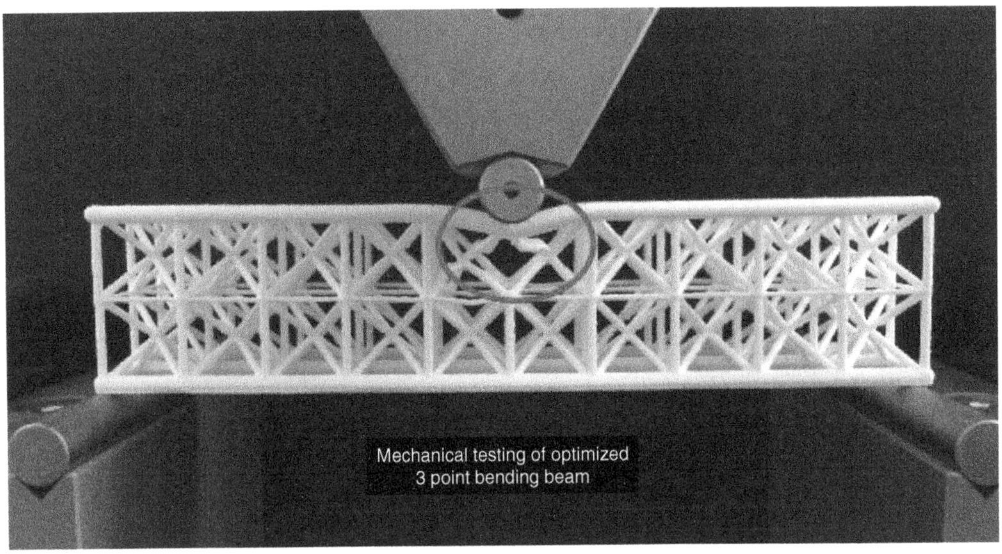

Figure 7.7 An example of the design validation.

7.3 Numerical Simulation of Lattice Structures

Simulation is important in the design process of lattice structures for three reasons. Firstly, designers need to decide the topology, size, and relative density of the lattice structure in order to satisfy the functional requirement. Simulation helps designers obtain the relation between the property and the geometry of the lattice structure. Secondly, the simulation result helps to improve the mechanical performance of lattice structures. The output from the simulation model, such as the stress and strain field, can help designers properly distribute the material in the lattice structure to optimize the design result. Thirdly, simulation can predict where and in which condition the failure could happen in the lattice structure. It helps designers to validate the design result. In the last few decades, the homogenization method and FEM are widely used in the simulation of the mechanical property of lattice structures.

7.3.1 Homogenization Method

The homogenization usually refers to a way to replace the composite with a kind of equivalent material model, which can overcome the difficulty in the analysis of the boundary value problem with high heterogeneities [35]. The mathematical theory of homogenization has developed in the 1970s [36, 37]. It is used to obtain the effective properties of homogenized material for periodic heterogeneous continuous media in many physical and engineering applications. It consists of two steps. Firstly, the local problem based on a unit cell is solved to get the effective material properties. Secondly, the overall problem is solved by substituting periodic material to the homogenized material with equivalent properties.

From a mathematical point of view, the theory of homogenization is a limit theory that uses the asymptotic expansion to substitute the rapidly oscillating coefficients in the differential equation with coefficients that are constant or slowly varying. The assumption of periodicity is used in this theory, and the solution of the substituted differential equation is close to the solution of the initial equation. It can be concluded that only the periodic and homogeneous lattice structure is suitable for the homogenization method. The computational cost of this method is low since the problem is simplified by the assumption of periodicity. A unit cell with periodic boundary conditions is used in the homogenization method to calculate the effective property of the lattice structure. The effective property can be assigned to a solid material, which is equivalent to the property of the lattice structure.

The lattice structures that can be analyzed by the homogenization method should have the following property:

$$G(\boldsymbol{x} + \boldsymbol{NY}) = G(\boldsymbol{x}) \tag{7.1}$$

where $\boldsymbol{x} = (x_1, x_2, x_3)$ is the position vector of the point, $\boldsymbol{N}$ is a 3×3 diagonal matrix:

$$\boldsymbol{N} = \begin{bmatrix} n_1 & 0 & 0 \\ 0 & n_2 & 0 \\ 0 & 0 & n_3 \end{bmatrix} \tag{7.2}$$

where n_1, n_2, and n_3 are arbitrary integer numbers, and $\boldsymbol{Y} = \lfloor y_1, y_2, y_3 \rfloor^T$ is a constant vector that determines the period of the structure. G can be a scalar or vectorial or even tensorial function of the position vector $\boldsymbol{x}$. For example, the constitutive relation in a unit cell is:

$$\sigma_{pq} = E_{pqrs}\,\varepsilon_{rs} \tag{7.3}$$

where σ_{pq} and ε_{rs} are the stress and strain tensors. E_{pqrs} is the locally varying stiffness tensor, which is a periodic function of the spatial coordinate $\boldsymbol{x}$, such that

$$E_{pqrs}(\boldsymbol{x} + \boldsymbol{NY}) = E_{pqrs}(\boldsymbol{x}) \tag{7.4}$$

After getting the local varying stiffness tensor, the macroscopic elasticity tensor E^H_{ijkl} can be computed as

$$E^H_{ijkl} = \frac{1}{|V|}\int_V E_{pqrs}\left(\varepsilon^{0(ij)}_{pq} - \varepsilon^{(ij)}_{pq}\right)\left(\varepsilon^{0(ij)}_{rs} - \varepsilon^{(ij)}_{rs}\right)dV \tag{7.5}$$

where $|V|$ denotes the volume of the unit cell, $\varepsilon^{0(ij)}_{pq}$ are prescribed macroscopic strain fields, while the locally varying strain fields $\varepsilon^{(ij)}_{pq}$ are defined as:

$$\varepsilon^{(ij)}_{pq} = \varepsilon_{pq}\left(\mathcal{X}^{ij}\right) = \frac{1}{2}\left(\mathcal{X}^{ij}_{p,q} + \mathcal{X}^{ij}_{q,p}\right) \tag{7.6}$$

based on the displacement fields $\mathcal{X}^{kl}$ found by solving the elasticity equations with a prescribed macroscopic strain

$$\int_V E_{ijpq}\,\varepsilon_{ij}(\mu)\varepsilon_{pq}\left(\mathcal{X}^{kl}\right)dV = \int_V E_{ijpq}\,\varepsilon_{ij}(\mu)\varepsilon^{0(kl)}_{pq}\,dV \;\forall \mu \in V \tag{7.7}$$

where μ is a virtual displacement field. For most practical problems, the homogenization is performed numerically by discretizing and solving Equation 7.7 using FEM.

The homogenization method can obtain the effective constitutive matrix of the lattice structure. But it is difficult to figure out the effective elastic modulus of the lattice structure. It is known that the lattice structure exhibit anisotropic material properties [38]. Although the effective constitutive matrix can tell the anisotropy, it is not clear in which direction the structure is strong and in which direction the structure is weak. To visually display the anisotropy of the lattice structure, the post-processing of the homogenization result is needed. One way to visualize the anisotropy of the lattice structure is to generate the Young's modulus surface. The Young's modulus surface is the Young's modulus of the lattice structure in the Cartesian coordinate. The vector from the origin to the point on the surface indicates the direction and the magnitude of the Young's modulus. It shows the value of the Young's modulus of the lattice structure in all the directions. If the material property is isotropic, the Young's modulus surface will be a spherical surface. Seven topologies shown in Figure 7.8 are analyzed by the homogenization model and the post-processing code. It should be noted that to achieve the same relative density, the structure thickness of each lattice topology is different. For example, Table 7.1 shows the strut thickness for unit cells with different lattice topologies to achieve 10, 30, and 50% relative densities. The effective Young's modulus surfaces of those lattice topologies with 10% relative density are shown in Figure 7.9.

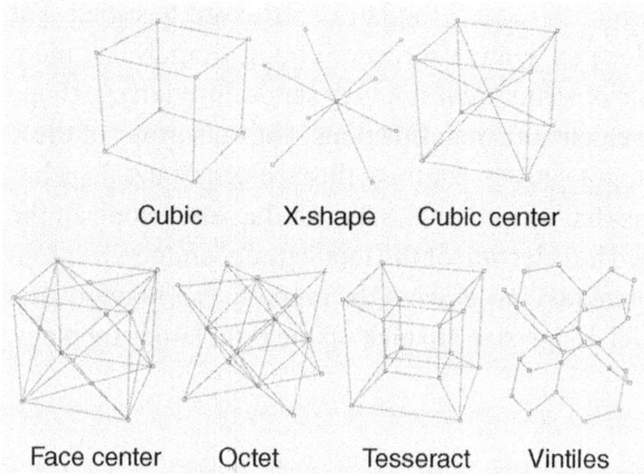

Figure 7.8 Example of lattice topologies.

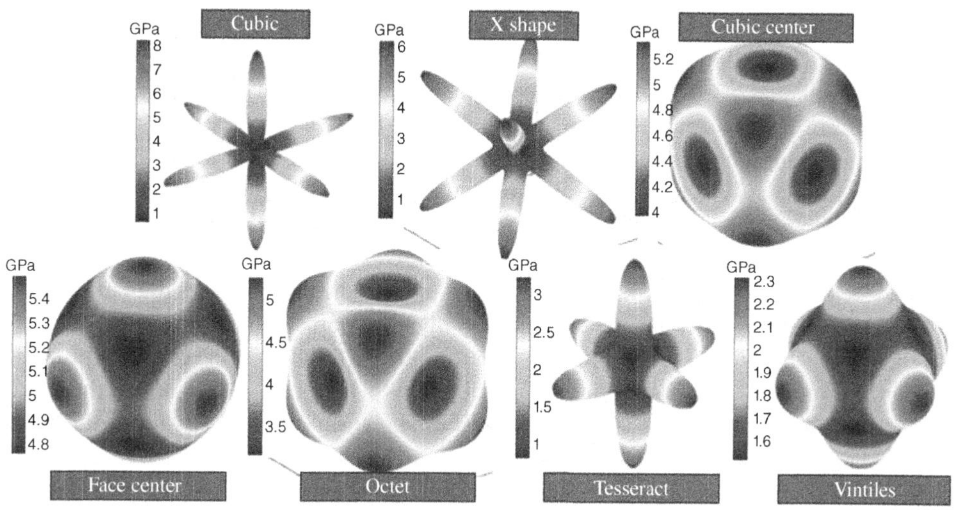

Figure 7.9 The Young's modulus surface of different lattice topologies.

Table 7.1 The parameters used to generate lattice structures with different topologies and relative densities.

	Cubic	X-shape	Cubic center	Face center	Octet	Tesseract	Vintiles
10%	0.113	0.074	0.063	0.058	0.045	0.054	0.065
30%	0.208	0.134	0.114	0.109	0.087	0.100	0.128
50%	0.287	0.187	0.157	0.150	0.120	0.138	0.177

The homogenization method can tell us in which direction the lattice structure is strong. For example, the cubic topology exhibits high modulus in x, y, and z directions but extremely low modulus in all other directions. It means that this type of structure is not very stable. Similarly, x-shape structure is strong in the diagonal direction but weak in all other directions. The anisotropy of these two topologies is significantly higher than other topologies. In contrast, the face-center topology has a nearly spherical modulus surface, which means that its modulus is almost the same along all the directions. It is very close to an isotropic property. The selection of the topology depends on its property. The Young's modulus surface can give us a general view of what the property is. Therefore, the homogenization method is often used in the initial design stage to pick up the right lattice topology.

7.3.2 Finite Element Method

FEM is a numerical method to solve real-world problems that involve complicated physics, geometry, and boundary conditions [39]. In FEM, a given domain is viewed as a collection of subdomains, and over each subdomain the governing equation is approximated by any of the variational methods. This is because it is easier to represent a complicated function as a collection of simple polynomials in the subdomains. There are two ways to simulate the mechanical property of lattice structures by FEM. One is the solid element model, as shown in Figure 7.10a; the other is the beam element model, as shown in Figure 7.10b. The solid element model is to divide the lattice structure into tetrahedrons or hexahedrons. Because the geometry of lattice structures is generally complicated, the solid element model is computationally heavy. The beam element model discretizes the lattice structure into struts. Each lattice strut is modeled with several beam elements. The quantity of beam elements to model a lattice structure is much less than using solid elements.

Different from the homogenization method, FEM has the capability of simulating both the homogeneous and heterogeneous lattice structures as well as periodic and nonperiodic lattice structures. However, the computational costs are relatively high compared to homogenization methods because the whole structure is taken into consideration. Even though the beam element model can reduce the computational cost, it has some limitations. The most obvious one is that beam elements cannot reflect the feature of the joint area as shown in Figure 7.11a–b. Furthermore, for the solid–lattice hybrid structures, an example of which is shown in Figure 7.11c, beam elements cannot mesh the whole structure. Solid elements must be used to model such a structure. Therefore, it is important to find a proper modeling method to simulate the mechanical properties of lattice structures, especially heterogeneous lattice and solid–lattice hybrid structures.

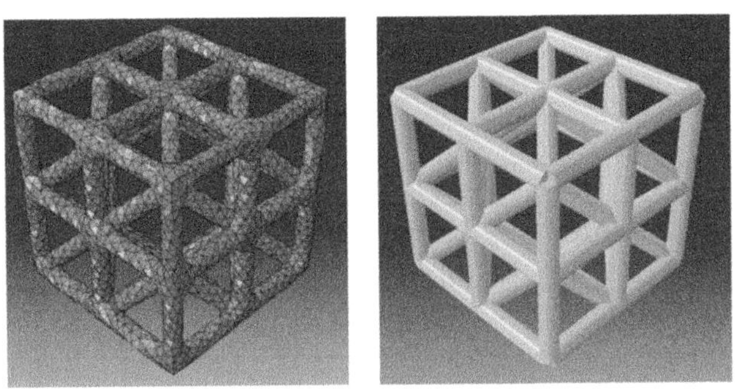

(a) (b)

Figure 7.10 Examples of the FEM models. (a) Solid element model and (b) beam element model.

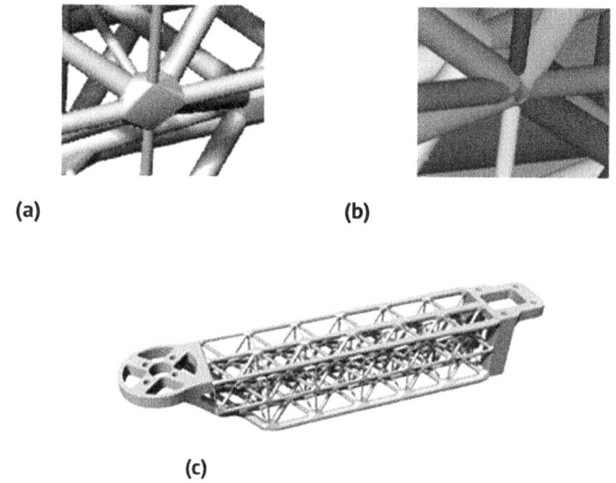

Figure 7.11 Challenges for beam element model. (a) Geometrial model of the joint, (b) beam element model for the joint, and (c) solid–lattice hybrid structure.

7.4 Lattice Structure Optimization Approaches

The most significant advantage of lattice structure is that the material distribution can be manipulated to achieve specific mechanical requirements. One of the most popular applications of lattice structure is to design lightweight, high stiffness-to-weight ratio parts. Two different optimization methods will be introduced in this section. They were both developed based on topology optimization theories to optimize the stiffness of the lattice structure by constraining the total weight.

7.4.1 Solid Isotropic Material with Penalization (SIMP)

The Solid Isotropic Material with Penalization (SIMP) method was developed in 1990s to optimize the topology with a volume constraint [40]. Recently, this method has been extended to optimize the density distribution of lattice structures [41]. Topology optimization assumes the stiffness of the material is related to the relative density of the finite element. The relative density is the design variable. Similarly, the optimization of lattice structure is to control the relative density distribution. The difference is that, in topology optimization, the stiffness constitutive matrix can be directly derived by a penalized equation. But the relationship between the relative density and the stiffness of the lattice structure has to be determined by the homogenization method.

The workflow of the SIMP method is shown in Figure 7.12. Homogenization method will be used to pre-calculate the effective property of the lattice structure with density range from 0 to 1. An interpolation function will be established by the data obtained from the homogenization method. The interpolation function will be used to calculate the corresponding property of the lattice unit cells in the structure. In each optimization iteration, the relative density will be updated, and the interpolation function will be called again to update the property of lattice unit cells. The loop will stop until the change of the relative density is smaller than a threshold.

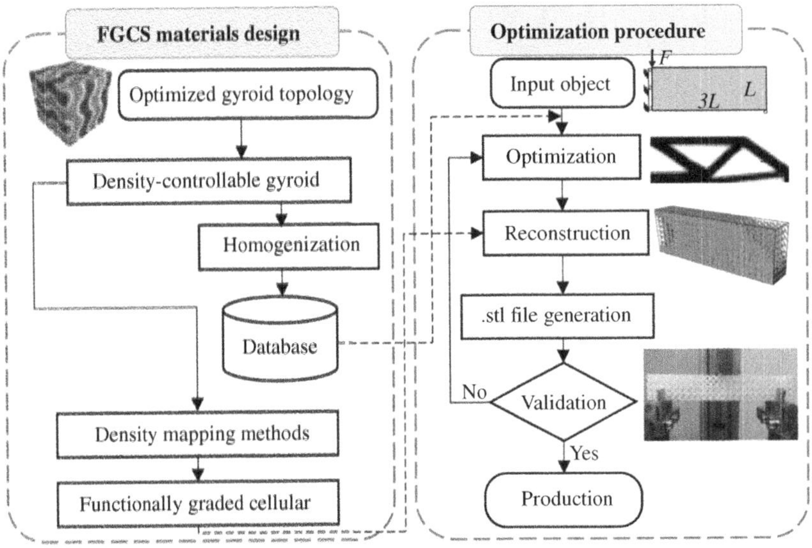

Figure 7.12 The workflow of the SIMP method. *Source*: [41]/with permission of Elsevier.

The objective function used here is to identify the optimal distribution of material density, subjected to target volume constraint, to minimize structural compliance. The mathematical representation of the optimization problem is:

$$\min c(\rho_r) = \frac{1}{2}\boldsymbol{u}^T\boldsymbol{K}(\rho_r)\boldsymbol{u} \tag{7.8}$$

$$s.t.\boldsymbol{f} = \boldsymbol{K}(\rho_r)\boldsymbol{u}$$

$$\boldsymbol{C}_G = \boldsymbol{C}_G(\rho_r)$$

$$V^* = \sum_{i=1}^{n} v_i \rho_r$$

$$0 < \rho_{\min} < \rho_r < \rho_{\max} < 1$$

where structural compliance $c(\rho_r)$ is the objective function, $\boldsymbol{K}$ is the global stiffness matrix, $\boldsymbol{f}$ is the external load vector, and $\boldsymbol{u}$ is the displacement vector. The first constraint is the equilibrium equation. The second constraint is the interpolation function of the property vs. relative density. The third constraint limits the total design volume to V^*. The final constraint ensures that the relative density of the structure remains within specified minimum and maximum limits. In standard topology optimization, the second constraint often employs a fictitious elastic scaling law to push design variables toward either void or solid, but this has no direct physical significance. In contrast, the proposed approach for cellular structure design replaces this fictitious constraint with a homogenization-based formulation that captures the actual mechanical properties of the cellular material as a function of its relative density. This allows the optimization to achieve intermediate densities that correspond to real, physically meaningful configurations of the cellular structure.

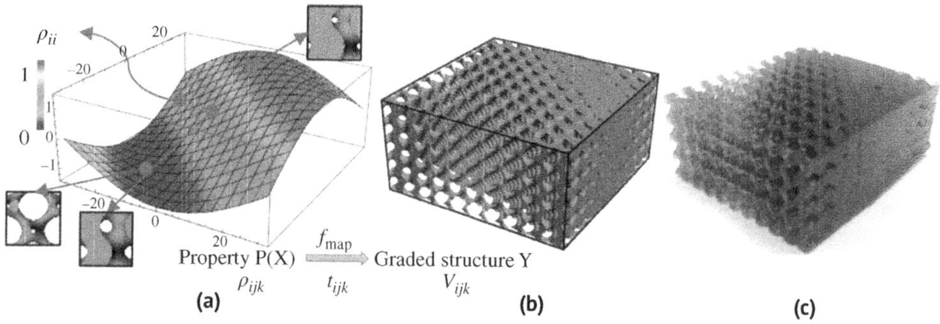

Figure 7.13 Functional graded structures controlled by the density distribution. *Source*: [41]/with permission of Elsevier.

This method is usually used for function-based lattice structures such as gyroid structures and other triply periodic minimal surface (TPMS) structures. The reason is that the generation of the lattice structure can be directly controlled by the density field, as shown in Figure 7.13. The density value is from 0 to 1. The function-based lattice structure can have a continuous density variation across the design space.

7.4.2 Bidirectional Evolutionary Structural Optimization

The bidirectional evolutionary structural optimization (BESO) method is a finite element-based topology optimization technique in which inefficient material is iteratively removed from a structure while simultaneously adding material in regions that contribute most to structural performance. Unlike the original evolutionary structural optimization (ESO) method, which only removes material, BESO improves both the efficiency and effectiveness of the optimization process. This method was first proposed by Querin et al. [42, 43] to enhance the optimization results and speed of the ESO algorithm. Later, a modified version of the BESO algorithm was presented by Huang and Xie [44] to address issues such as non-convergence and mesh dependency encountered in early implementations.

In the context of lattice structures, a further modification of the BESO algorithm has been developed to optimize the thickness of individual lattice struts [7]. Instead of directly removing or adding elements as in the conventional BESO method, the thickness of each strut is iteratively updated to redistribute material throughout the design space. This approach mimics the bone remodeling process described by Wolff's law [45]. The strut thickness is reduced in low-stress areas and is increased in the high-stress areas. The volume of material removed is equal to material added, which keeps the total volume of the lattice structure unchanged. The mathematical representation of the proposed optimization problem for lattice structure can be expressed as:

$$\text{To Find}: t_k, k = 1,2\ldots,n \tag{7.9}$$

$$\text{Minimize}: PI = \frac{\sum_{i=1}^{n} \sigma_{VM_i} V_i}{FL}$$

$$S.T.: Ku - P = 0$$

$$\sum_{i=1}^{n} V_i \leq V_{\text{const}}$$

$$t_{\min} \leq t_k \leq t_{\max}$$

where t_k is the thickness of the k_{th} strut in lattice frame; n is the number of struts in the lattice frame; PI is the performance indicator of the structure proposed by Querin [32] and Steven [36] to measure how well the overall structure is performing against an idealized fully stressed design. σ_{VM_i} and V_i are the maximum von Mises stress and volume of i_{th} strut; F is a representational force and L is a reference length. These two parameters are problem specific and constant throughout the optimization process. K is the global stiffness matrix of lattice structure and u is the node displacement vector; P is the external load. V_{const} is the given volume of the lattice structure; $t_{\min}$ and $t_{\max}$ are the minimum thickness and maximum thickness of lattice struts, respectively. The general workflow of the algorithm is illustrated in Figure 7.14. The core concept of this method is to redistribute the material or volume of lattice struts based on the stress distribution obtained from FEA. A central feature of the algorithm is the volume relocation process: struts with maximum von Mises stress below a specified threshold have their volume reduced by a certain proportion, and the total removed material is then redistributed to struts according to their stress levels. Through repeated iterations of this process, the lattice structure gradually adapts to the applied external loads, improving its overall mechanical performance.

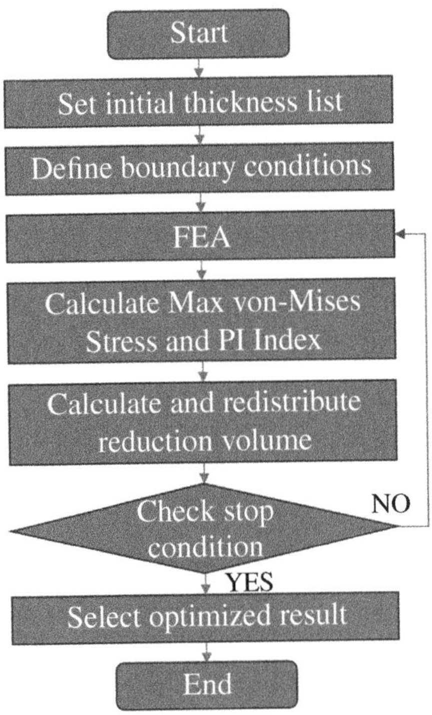

Figure 7.14 General working flow of BESO-based lattice optimization algorithm. *Source*: Adapted from [7].

7.5 Design Cases of Lattice Structures

In the end of this chapter, a case study is used to demonstrate the design method for lattice structures. Simulation analyses are conducted to evaluate the stiffness and linear buckling load of the solid–lattice hybrid structure. These results are compared with those of equivalent pure solid and pure lattice structures, all designed to have the same weight, in order to assess the performance benefits of the hybrid approach. In addition, physical experiments are carried out to validate the simulation results and confirm the improvements in mechanical behavior, including enhanced stiffness and load-carrying capacity, achieved by the hybrid design.

The boundary condition of the design case are illustrated in Figure 7.15a. The beam has dimensions of 200 mm × 60 mm × 30 mm, and a 100 N load is applied as a uniform line load along the center of the top surface. In the initial design stage, the solid space is defined by identifying the regions that has a high-density in the topology optimization result as shown in Figure 7.15b–c. The results indicate that the material density at the top-left and top-right corners is effectively zero, suggesting that these regions do not contribute to structural performance and can be removed. Subsequently, additional regions with low relative density within the interior of the design space are defined as the lattice space.

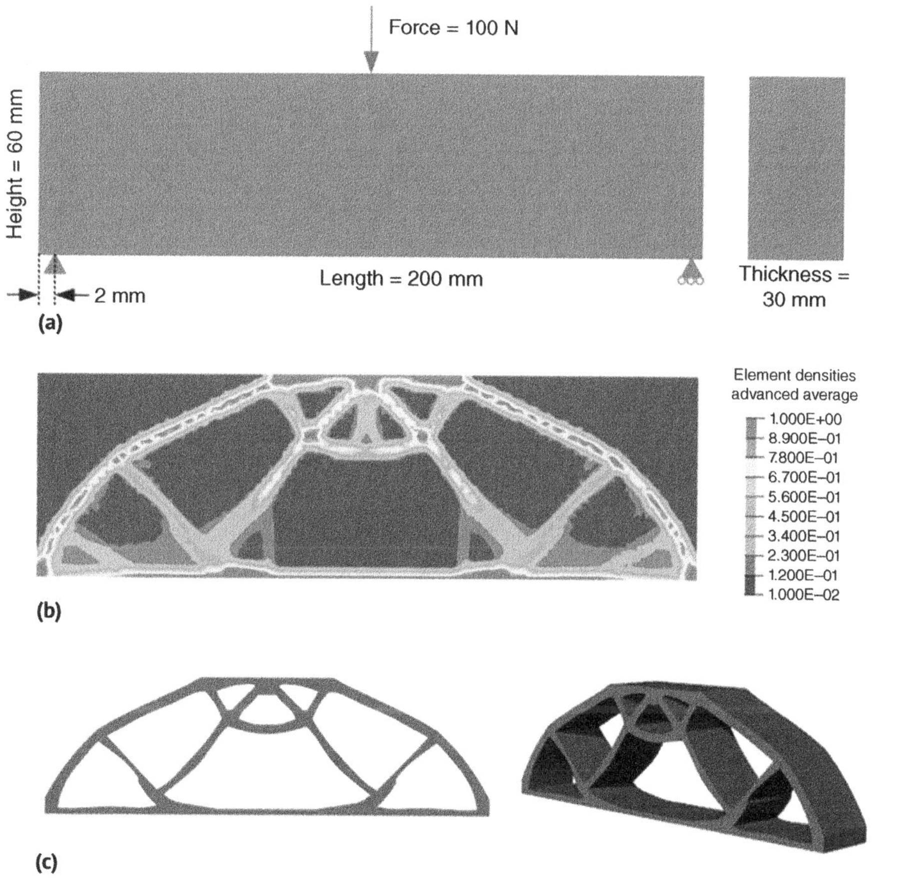

Figure 7.15 The topology optimization of the three-point bending beam. (a) Boundary conditions, (b) topology optimization result, and (c) solid part design. *Source*: [1]/with permission of Elsevier.

In this case study, the lattice wireframe is generated through a two-step process. In the first step, a cubic-center topology is used to fill all the empty regions within the solid part. The unit cell size is set to 10 mm in the top central region and 15 mm in the remaining areas. A simulation model combining the solid part and the lattice wireframe is then created to evaluate the structural performance. The optimization results reveal that in the lower central region, all the inclined struts reach the minimum allowable thickness, indicating that the cubic-center topology is redundant in this area. Consequently, the topology in this region is modified to a standard cubic configuration as shown in Figure 7.16a. In the second step, short and ineffective struts are removed, and the lattice wireframe is further refined using the BESO method. The updated lattice configuration obtained from this second optimization round is presented in Figure 7.16b.

The beam designed using the proposed hybrid method is compared with both a pure lattice structure and a pure solid structure, as shown in Figure 7.17. The pure lattice structure is optimized separately using the BESO method, while the pure solid structure is derived directly from the topology optimization result. All three designs have the same weight, allowing for a fair comparison. Their mechanical performance is evaluated in terms of stiffness and critical buckling load to assess the advantages of the hybrid approach over the conventional pure lattice or solid designs.

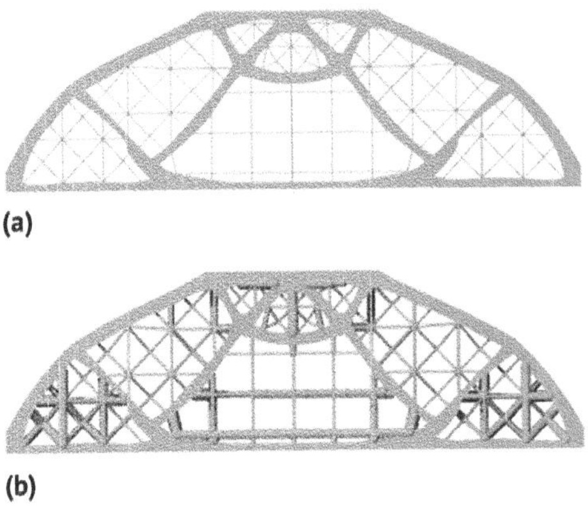

Figure 7.16 The generation of the lattice wireframe. (a) Modified lattice wireframe and (b) optimized lattice structure. *Source*: [1]/with permission of Elsevier.

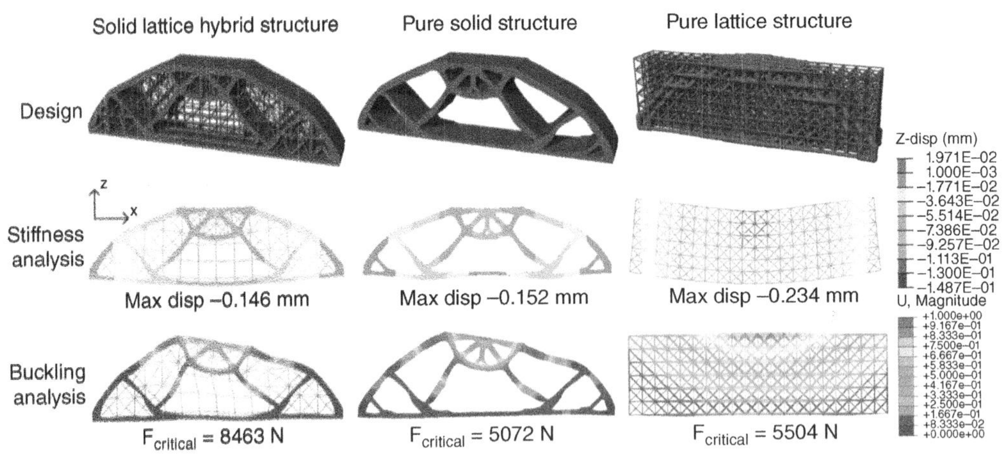

Figure 7.17 Three designs of the three-point bending beam and the comparison of the stiffness. *Source*: [1]/with permission of Elsevier.

The mechanical performance of the three designs was first evaluated through finite element analysis. Stiffness analysis, as shown in Figure 7.17, reveals that the hybrid structure exhibits the smallest displacement under load, indicating the highest stiffness among the three designs. The pure solid structure shows slightly larger displacements, while the pure lattice design exhibits the lowest stiffness. These differences can be attributed to the distribution of material: the hybrid structure incorporates lattice infill in intermediate-density regions identified by topology optimization, allowing it to reinforce areas that the pure solid structure cannot optimize further. In contrast, the pure lattice structure cannot achieve full material density in critical regions due to strut thickness limitations, limiting its stiffness despite the presence of intermediate-density regions.

Linear buckling analysis was also performed by applying a unit load at the top center of the beam, with the critical load calculated as the product of the unit load and the eigenvalue. The results indicate that the hybrid structure has the highest critical buckling load, demonstrating superior resistance to buckling. Compared to the hybrid design, the critical loads of the pure solid and pure lattice structures are approximately 40 and 35% lower, respectively. The pure solid structure is particularly vulnerable due to thin shell regions predicted by topology optimization, which experience high compressive stresses and initiate buckling. In the hybrid structure, lattice elements connect to these thin shells, providing additional support and delaying buckling, which is a key advantage of the solid–lattice hybrid design.

To validate the simulation results, all three designs were fabricated using a Zortrax M200 3D printer with acrylonitrile butadiene styrene (ABS) as the printing material. The fabrication process was standardized across all samples. Physical testing was conducted on a Test Resources 313 universal testing machine at a loading speed of 1 mm/min, as shown in Figure 7.18. The load–displacement responses, presented in Figure 7.19, confirm that the hybrid structure achieves the highest stiffness and ultimate strength, consistent with the FEA predictions. The pure solid structure began to buckle at a displacement of 6 mm, while the hybrid structure remained stable under the same conditions. The failure patterns, shown in Figure 7.20, further illustrate these differences: the pure solid structure failed by buckling in the regions predicted by simulation, whereas the hybrid structure withstood higher loads and ultimately fractured before any buckling occurred.

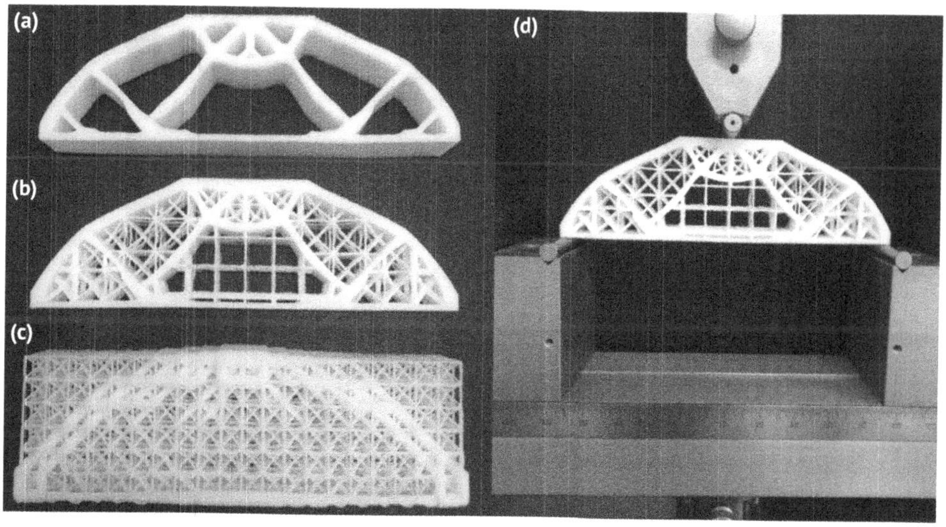

Figure 7.18 The structures fabricated by the FDM process and the experiment setup. (a) Pure solid, (b) hybrid, (c) pure lattice, and (d) experiment setup. *Source*: [1]/with permission of Elsevier.

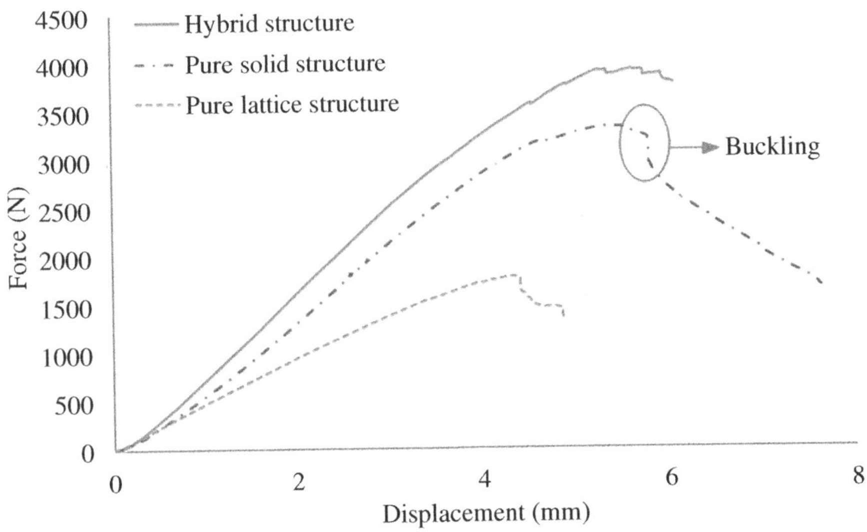

Figure 7.19 The load–displacement curves of the three structures. *Source*: [1]/with permission of Elsevier.

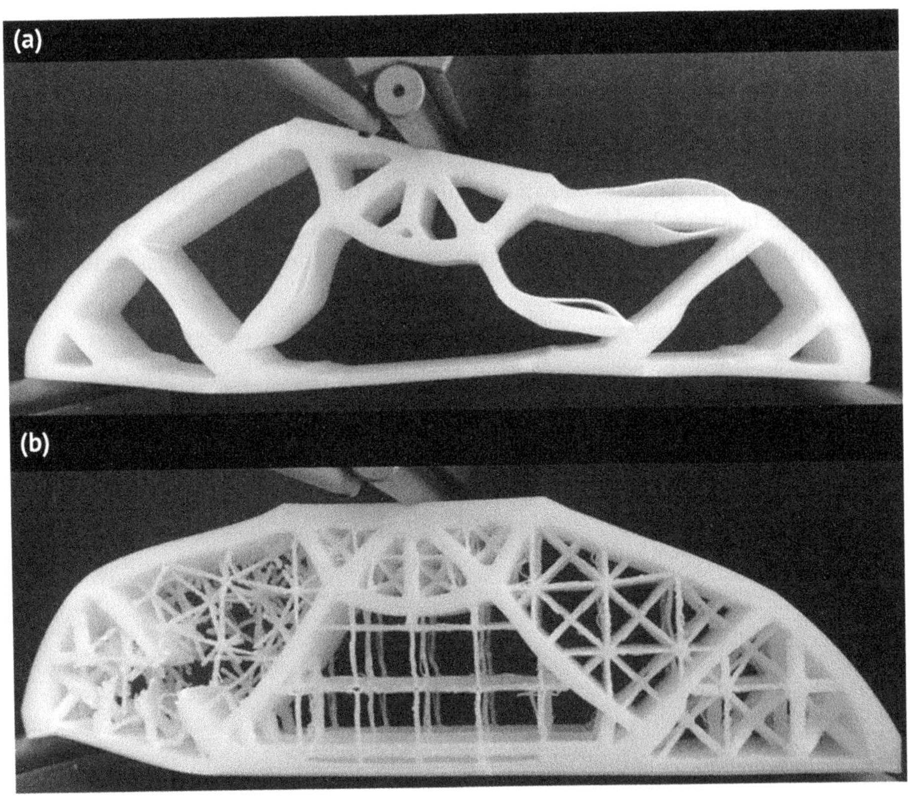

Figure 7.20 Failure of the (a) pure solid structure and (b) solid–lattice hybrid structure. *Source*: [1]/with permission of Elsevier.

These results demonstrate that integrating lattice infill into the solid design not only enhances stiffness but also increases the critical buckling load, effectively reinforcing thin-walled regions and improving the overall mechanical performance of the component. Both simulation and experimental evidence highlight the advantages of the solid–lattice hybrid approach over conventional pure solid or lattice structures.

7.6 Bio-inspired Lattice Design

Multifunctional bio-inspired design (MBID) offers complementary value by helping designers systematically navigate and balance function-driven ideation with manufacturing constraints. MBID facilitates early-stage conceptualization of multifunctional components through functional decomposition, biomimetic analogy, and structure–function mapping. Lattice structures were initially found in nature. MBID offers a conceptual-level design idealization, where lattice structures serve as design candidates. Voronoi lattice structure, a type of random lattice, has been widely used in bio-inspired design. For example, hip bones and dragonfly wings are constructed by 3D and 2D Voronoi cells, as shown in Figure 7.21. Li et al. [46] proposed a bio-inspired variable-periodic Voronoi tessellation method to optimize the lattice structure. Although lattice structures in nature can help designers initiate the topology in the conceptual design stage, a detailed optimization algorithm is still needed in the detailed design stage.

Beyond hip bones and dragonfly wings, a wide range of biological systems exhibit lattice-like architectures that have inspired advanced engineering applications. For instance, the internal structure of cuttlebone demonstrates a highly porous, yet mechanically robust, layered lattice system ideal for lightweight structural applications. Similarly, the beak of the toucan features a foam-like trabecular core encased in a dense outer shell, optimizing both strength and energy absorption. In plant biology, the cellular arrangement of sunflower stems exhibits hierarchical lattice-like networks that combine stiffness with flexibility. These natural inspirations have led to the development of functionally graded and hierarchical lattice structures in AM, where mechanical performance can be tailored locally through controlled geometry variations. Such examples highlight how MBID not only facilitates initial form generation but also provides a biological rationale for performance-driven customization in complex engineering systems.

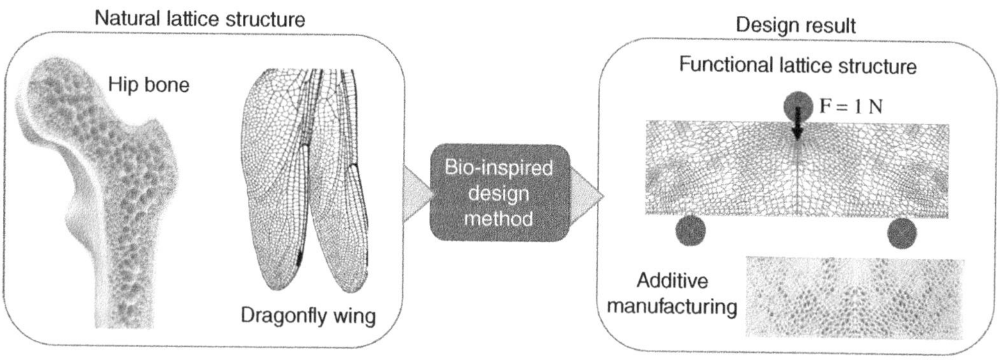

Figure 7.21 Voronoi structures in the nature and bio-inspired design approach for Voronoi lattice structure.

7.7 Conclusion

In this chapter, the design methodology for lattice structures made by AM was introduced. The three design stages of lattice structures are supported by simulation methods, including the homogenization method and FEM. It is important to understand the advantages and disadvantages of the simulation methods for lattice structures. Proper simulation methods can facilitate the design and optimization of lattice structures in different design stages. The optimization methods for lattice structures were originally developed from the topology optimization methods such as the SIMP and BESO methods. These methods were modified to consider the cellular characteristics of lattice structure. A design case study for lattice structure was introduced in this chapter to illustrate the application of the design methodology for lattice structures. In addition, bio-inspired design approaches for lattice structures are introduced. Overall, the DfAM for lattice structures is to exploit the geometrical complexity provided by AM. In Chapter 8, the focus of DfAM will shift from geometrical complexity to material complexity. Advanced topics about digital materials and emerging paradigms in DfAM will be explored.

References

1 Dong, G., Tang, Y., and Zhao, Y.F., *A survey of modeling of lattice structures fabricated by additive manufacturing*. Journal of Mechanical Design, 2017. **139**(10): p. 100906-100906-13.

2 Evans, A.G., Hutchinson, J., and Ashby, M., *Multifunctionality of cellular metal systems*. Progress in Materials Science, 1998. **43**(3): pp. 171–221.

3 Deshpande, V.S., Fleck, N.A., and Ashby, M.F., *Effective properties of the octet-truss lattice material*. Journal of the Mechanics and Physics of Solids, 2001. **49**(8): pp. 1747–1769.

4 Lehman, J. and Lakes, R., *Stiff lattices with zero thermal expansion and enhanced stiffness via rib cross section optimization*. International Journal of Mechanics and Materials in Design, 2013. **9**(3): pp. 213–225.

5 Yang, L., et al., *Non-stochastic Ti-6Al-4V foam structures with negative Poisson's ratio*. Materials Science and Engineering A, 2012. **558**: pp. 579–585.

6 Zhang, Q., et al., *Ultralight X-type lattice sandwich structure (I): concept, fabrication and experimental characterization*. Science in China Series E: Technological Sciences, 2009. **52**(8): pp. 2147–2154.

7 Tang, Y., Kurtz, A., and Zhao, Y.F., *Bidirectional Evolutionary Structural Optimization (BESO) based design method for lattice structure to be fabricated by additive manufacturing*. CAD Computer Aided Design, 2015. **69**: pp. 91–101.

8 Rosen, D.W., *Computer-aided design for additive manufacturing of cellular structures*. Computer-Aided Design and Applications, 2007. **4**(5): pp. 585–594.

9 Ozdemir, Z., et al., *Energy absorption in lattice structures in dynamics: experiments*. International Journal of Impact Engineering, 2016. **89**: pp. 49–61.

10 Lehman, J. and Lakes, R.S., *Stiff, strong, zero thermal expansion lattices via material hierarchy*. Composite Structures, 2014. **107**: pp. 654–663.

11 Brooks, H. and Brigden, K., *Design of conformal cooling layers with self-supporting lattices for additively manufactured tooling*. Additive Manufacturing, 2016. **11**: pp. 16–22.

12 Arabnejad Khanoki, S. and Pasini, D., *Fatigue design of a mechanically biocompatible lattice for a proof-of-concept femoral stem*. Journal of the Mechanical Behavior of Biomedical Materials, 2013. **22**: pp. 65–83.

13 Masood, S.H., Singh, J.P., and Morsi, Y., *The design and manufacturing of porous scaffolds for tissue engineering using rapid prototyping*. International Journal of Advanced Manufacturing Technology, 2005. **27**(3–4): pp. 415–420.

14 Cheah, C.M., et al., *Development of a tissue engineering scaffold structure library for rapid prototyping. Part 1: investigation and classification*. The International Journal of Advanced Manufacturing Technology, 2003. **21**(4): pp. 291–301.

15 Lin, Y., et al., *Combination of polyetherketoneketone scaffold and human mesenchymal stem cells from temporomandibular joint synovial fluid enhances bone regeneration*. Scientific Reports, 2019. **9**(1): p. 472.

16 Gibson, L.J. and Ashby, M.F., *Cellular solids: structure and properties*, Second Edition. 1997, Cambridge University Press. pp. 1–510.

17 Wang, H., Chen, Y., and Rosen, D.W., A hybrid geometric modeling method for large scale conformal cellular structures. in *ASME 2005 International Design Engineering Technical Conferences and Computers and Information in Engineering Conference*. 2005, 47403: pp. 421–427.

18 Wang, H. and Rosen, D.W., Parametric modeling method for truss structures. *ASME 2002 International Design Engineering Technical Conferences and Computers and Information in Engineering Conference*. 2024. 2002, 36215: pp. 759–767.

19 Wang, H.V., *A unit cell approach for lightweight structure and compliant mechanism*. 2005, Citeseer.

20 Nguyen, J., et al., Conformal lattice structure design and fabrication. in Solid Freeform Fabrication Symposium, Austin, TX. 2012.

21 Tao, W. and Leu, M.C., Design of lattice structure for additive manufacturing. in *International Symposium on Flexible Automation, ISFA 2016*. 2016.

22 Tang, Y., et al., *Lattice structure design and optimization with additive manufacturing constraints*. IEEE Transactions on Automation Science and Engineering, 2017. 99: pp. 1–17.

23 Melchels, F.P.W., Feijen, J., and Grijpma, D.W., *A review on stereolithography and its applications in biomedical engineering*. Biomaterials, 2010. **31**(24): pp. 6121–6130.

24 Rezaei, R., et al., Mechanical characterization and finite element modeling of polylactic acid BCC-Z cellular lattice structures fabricated by fused deposition modeling. in *Proceedings of the Institution of Mechanical Engineers, Part C: Journal of Mechanical Engineering Science*, 2016: p. 0954406215626941.

25 Naghieh, S., et al., *Numerical investigation of the mechanical properties of the additive manufactured bone scaffolds fabricated by FDM: The effect of layer penetration and post-heating*. Journal of the Mechanical Behavior of Biomedical Materials, 2016. **59**: pp. 241–250.

26 Tang, Y., et al., *Elastic modulus of 316 stainless steel lattice structure fabricated via binder jetting process*. Materials Science and Technology (United Kingdom), 2016. **32**(7): pp. 648–656.

27 Yan, C., et al., *Evaluations of cellular lattice structures manufactured using selective laser melting*. International Journal of Machine Tools and Manufacture, 2012. **62**: pp. 32–38.

28 Cerardi, A., et al., *Mechanical characterization of polyamide cellular structures fabricated using selective laser sintering technologies*. Materials and Design, 2013. **46**: pp. 910–915.

29 Murr, L.E., et al., Additive layered manufacturing of reticulated Ti-6Al-4V biomedical mesh structures by electron beam melting. in *25th Southern Biomedical Engineering Conference 2009*. 2009. Miami, FL.

30 Parthasarathy, J., et al., *Mechanical evaluation of porous titanium (Ti6Al4V) structures with electron beam melting (EBM)*. Journal of the Mechanical Behavior of Biomedical Materials, 2010. **3**(3): pp. 249–259.

31 Schaedler, T.A. and Carter, W.B., *Architected cellular materials*. Annual Review of Materials Research, 2016. pp. 187–210.

32 Mun, J., et al., *Indirect additive manufacturing based casting of a periodic 3D cellular metal - Flow simulation of molten aluminum alloy*. Journal of Manufacturing Processes, 2015. **17**: pp. 28–40.

33 Schaedler, T.A., et al., *Ultralight metallic microlattices*. Science, 2011. **334**(6058): pp. 962–965.

34 Cheung, K.C. and Gershenfeld, N., *Reversibly assembled cellular composite materials*. Science, 2013. **341**(6151): pp. 1219–1221.

35 Hassani, B. and Hinton, E., *A review of homogenization and topology optimization I—homogenization theory for media with periodic structure*. Computers & Structures, 1998. **69**(6): pp. 707–717.

36 Bensoussan, A., Lions, J.-L., and Papanicolaou, G., *Asymptotic analysis for periodic structures*. Vol. 374. 1978, American Mathematical Soc.

37 Cioranescu, D. and Paulin, J.S.J., *Homogenization in open sets with holes*. Journal of Mathematical Analysis and Applications, 1979. **71**(2): pp. 590–607.

38 Xu, S., et al., *Design of lattice structures with controlled anisotropy*. Materials and Design, 2016. **93**: pp. 443–447.

39 Reddy, J.N., *An introduction to the finite element method*. Vol. 2. 1993.

40 Sigmund, O., *Design of material structures using topology optimization*. 1994, Technical University of Denmark Lyngby.

41 Li, D., et al., *Optimal design and modeling of gyroid-based functionally graded cellular structures for additive manufacturing*. Computer-Aided Design, 2018. **104**: pp. 87–99.

42 Querin, O.M., Steven, G.P., and Xie, Y.M., *Evolutionary structural optimisation (ESO) using a bidirectional algorithm*. Engineering computations, 1998. **15**(8): pp. 1031–1048.

43 Young, V., et al., *3D and multiple load case bi-directional evolutionary structural optimization (BESO)*. Structural optimization, 1999. **18**(2): pp. 183–192.

44 Huang, X. and Xie, Y., *Convergent and mesh-independent solutions for the bi-directional evolutionary structural optimization method*. Finite elements in analysis and design, 2007. **43**(14): pp. 1039–1049.

45 Frost, H.M., *Wolff's Law and bone's structural adaptations to mechanical usage: an overview for clinicians*. The Angle Orthodontist, 1994. **64**(3): pp. 175–188.

46 Li, Z., Chu, S., and Wu, Z., *A novel bio-inspired design method for porous structures: Variable-periodic Voronoi tessellation*. Materials & Design, 2024. **243**: p. 113055.

8

Advanced Topics in DfAM: Digital Materials and Emerging Paradigms

Building upon the foundational and process-specific design principles in Chapters 3–7, this chapter explores the advanced opportunities enabled by digital material systems. Additive Manufacturing (AM) has four axes of complexity, which highlight the unique value propositions of AM compared to traditional manufacturing. In the previous chapters, DfAM topics have addressed the **geometric complexity** (Chapter 4 part consolidation, Chapter 6 topology optimization, and Chapter 7 lattice structure), **functional complexity** (Chapter 6 multifunctional optimization), and **hierarchical complexity** (Chapter 6 multi-scale structural design and Chapter 7 solid-lattice hybrid structure). The last complexity, **material complexity**, is defined by the use of different materials or material gradients in a single build or across builds. This chapter explores advanced topics in DfAM that leverages the material complexity provided by AM. Digital materials design approaches for AM will be introduced including multi-material design, design for voxel printing, and fiber-reinforced composite design.

8.1 Introduction to Digital Materials

Digital materials refer to a new category of materials, the composition of which can be digitally controlled and manufactured. Traditional materials are typically homogeneous and have a uniform composition, like plastics, alloys, and composites. Although uniform materials can exhibit excellent mechanical properties, they cannot satisfy complex design requirements. For example, soft robotics can change shapes by stimuli to achieve movements. One of the methods to fabricate such soft robotics is to have a functionally graded material that has different thermal expansion on different sides [1]. If the temperature increases, one side will expand more and cause a bending moment.

AM has accelerated the development of digital materials because several manufacturing technologies enabled the manufacturing of digital materials. Multi-material printing, voxel printing, and continuous fiber-reinforced polymer printing have been increasingly used to fabricate digital materials. However, the design of digital materials is difficult because existing design methodologies are developed for traditional materials. The focus is on the structure instead of the materials themselves, like topology optimization. The focus of digital materials design is on the multi-material distribution in the design space. In this chapter, the design methodologies developed for digital materials will be introduced.

8.2 Fabrication of Digital Materials

8.2.1 Multi-material AM

Multi-material AM refers to the process of creating objects or structures with the ability to incorporate multiple materials simultaneously. In multi-material AM, various materials with distinct properties, such as polymers, metals, ceramics, or composites (as shown in Figure 8.1) can be deposited layer by layer to create complex and heterogeneous structures [2, 3]. This enables the production of components with tailored properties, such as gradient materials [4], lattice structures [5], or objects with different material compositions in different regions. Multi-material AM offers several advantages, including the ability to integrate multiple functionalities or properties into a single part [6], enhance mechanical performance through material combinations [7], and create intricate internal structures [8]. Multi-material AM can be achieved by material jetting [9], material extrusion [10, 11], stereolithography (SLA) [12], and powder bed fusion (PBF).

Material extrusion, such as fused deposition modeling (FDM) and direct ink writing (DIW), is the most common and affordable process to fabricate multi-material structures. In FDM-based multi-material printing, distinct filaments of different materials are carefully orchestrated to merge into a single printed object. This orchestration is achieved by equipping the 3D printer with multiple extruders, each extruder feeding a different material. As the object takes shape layer by layer, the printer can switch between extruders, enabling the incorporation of various materials. This capability opens the door to functional gradients, composite structures, and assemblies of parts with varying mechanical, thermal, or even electrical properties. From automotive components designed for specific stress distributions to consumer products marrying rigid and flexible sections seamlessly, FDM's multi-material capabilities have expanded the horizons of design and engineering. DIW's multi-material capabilities are realized through carefully designed printing heads that accommodate different materials. By manipulating the deposition parameters and material combinations, DIW can generate intricate structures with graded mechanical properties or custom-designed composite characteristics. From tissue engineering applications that combine biocompatible and supportive materials to electronic devices featuring conductive and insulating elements, DIW showcases the potential of multi-material printing in sectors that demand tailored material functionalities.

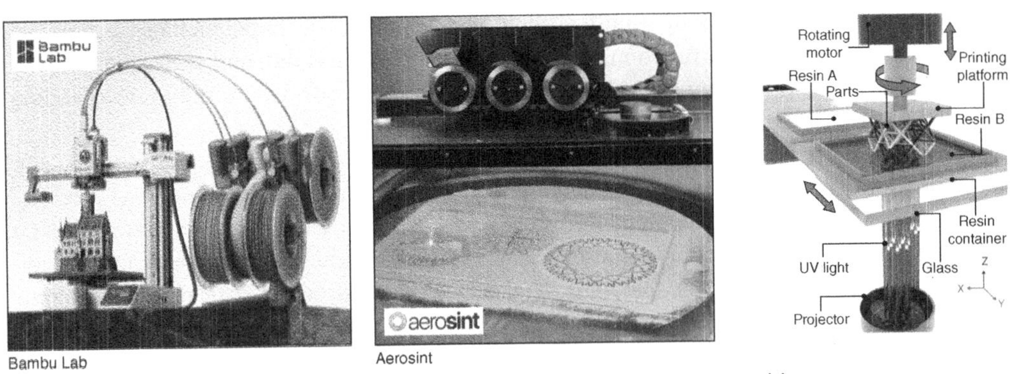

(a) (b) (c)

Figure 8.1 Multi-material AM technologies. (a) Material extrusion, (b) powder bed fusion, and (c) stereolithography. *Source*: (a) https://bambulab.com/en-us/a1-mini; (b) https://aerosint.com/; (c) https://www.nature.com/articles/s41467-022-35622-6.

Metal PBF is a cutting-edge AM technique that enables the creation of complex metal parts with exceptional precision and control. As industries demand higher levels of customization, improved performance, and intricate designs in metal components, the evolution of multi-material PBF has emerged as a significant breakthrough [13]. This innovative approach not only leverages the inherent benefits of PBF, such as high resolution and material efficiency, but also expands the potential applications by allowing the combination of different metals within a single printed object. The breakthrough of multi-material PBF lies in the integration of different metal powders within a single build. This is achieved through a combination of powder deposition strategies and intricate beam control mechanisms. Multi-material metal PBF opens up remarkable possibilities across industries, including aerospace, automotive, and medical sectors. It enables the creation of objects with gradients, engineered material combinations, and intricate internal structures that were previously challenging or impossible to manufacture using traditional methods. As researchers and engineers continue to refine the technology, multi-material metal PBF stands at the forefront of AM innovation, offering new dimensions of customization and performance in the world of metal parts fabrication.

SLA is a powerful AM technology that utilizes light and photosensitive resin to build 3D multi-material objects layer by layer. In this process, a reservoir of liquid resin containing different material formulations is selectively cured by a digital projector emitting ultraviolet (UV) light. Multi-material SLA printing has challenges and difficulties due to the complexities involved in coordinating material compatibility, curing dynamics, software, and control. Generally, multiple reservoirs are filled with multiple types of resins. The light cures one resin each time. Before switching to the next type of resin, a cleaning mechanism is required to remove the uncured resin. The cleaning mechanism can be realized by mechanical brushes, alcohol wash, centrifugal force [14], etc. By intricately coordinating the deposition of different resin formulations at precise locations on each layer, multi-material SLA realizes complex objects with varying material behaviors. This innovation allows for the seamless integration of rigid, flexible, transparent, and other specialized materials within a single print. As a result, industries ranging from aerospace to consumer goods can explore the unlimited design potential that arises from this multi-material approach, promising functional complexity and enhanced performance across diverse applications.

8.2.2 Voxel Printing

Material jetting is the most mature and flexible process in the fabrication of multi-material structures, as it can control the material voxel by voxel [15]. This technique offers exceptional accuracy, fine detail, and the ability to produce multi-material and multi-color objects, making it particularly attractive for applications that demand intricate geometries and customized aesthetics.

Material jetting involves the deposition of droplets of liquid photopolymer material onto a build platform. These droplets are precisely placed by an array of inkjet printheads guided by a digital design. The liquid droplets are rapidly cured using UV light, solidifying each layer of the object. The build platform is then lowered, and a new layer of material is deposited and cured on top of the previous one. This layer by layer process continues until the object is fully formed. Voxel printing is a crucial aspect of the material jetting process that enables precise control over the deposition of materials and properties at a very fine scale. A "voxel" is a three-dimensional equivalent of a pixel in two-dimensional imagery. It represents a tiny volumetric element within a 3D space. Voxel printing is the technique of manipulating and controlling the properties of individual voxels during the material jetting process. This level of control allows for the creation of complex multi-material, multi-color, and gradient objects with incredible precision. Examples of voxel printing are shown in Figure 8.2, which demonstrates the flexibility of material control in the material jetting process.

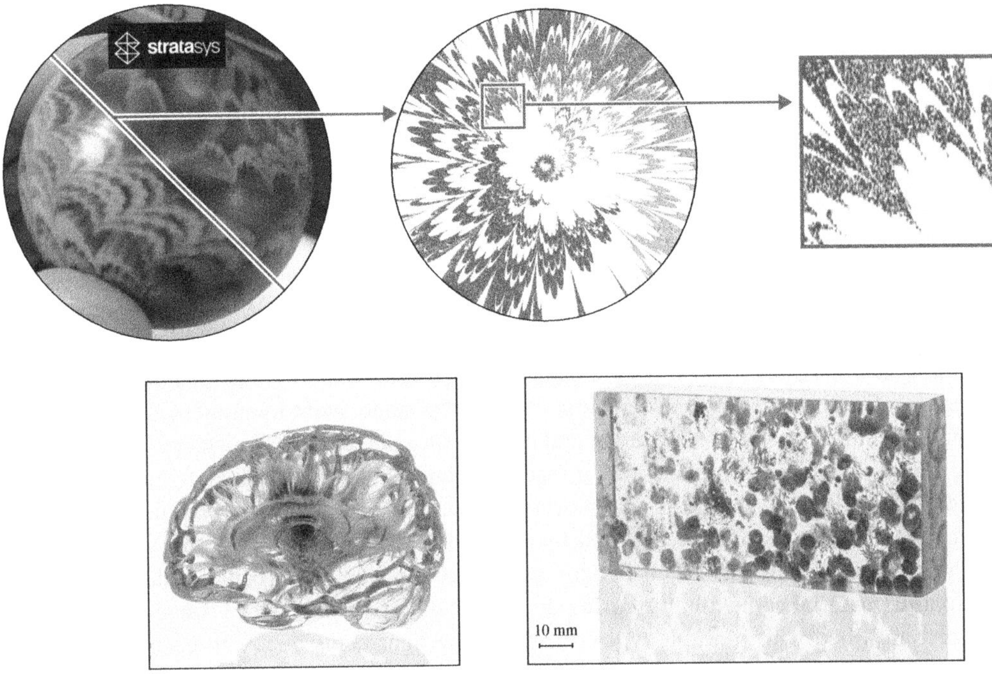

Figure 8.2 Examples of voxel-printed objects. *Source*: Stratasys Ltd. / https://help.grabcad.com/article/230-guide-to-voxel-printing?locale=en / last accessed on Oct 10, 2025; [60]/with permission of American Association for the Advancement of Science (AAAS).

Voxel printing is an advanced form of AM that marries digital design with precise material deposition, resulting in objects of unparalleled complexity and customization. As material jetting technology advances, the potential to control individual voxels opens doors to creative expression, functional innovation, and the production of highly customized and intricate objects.

8.2.3 Fiber-reinforced Polymer Printing

Additively manufactured fiber-reinforced polymers can be classified into continuous fiber and short fiber composites [16]. 3D printing of polymer composites with nanoparticles (e.g., carbon black, carbon nanotubes) and short fibers have been demonstrated using PBF, SLA [1, 17–19], and inkjet [20, 21]. However, the direct incorporation of continuous fiber is challenging for these processes. Recently, the challenges in AM of continuous fiber reinforced composites have been overcome by several processes including FDM [22–27] and DIW [28–33], as shown in Figure 8.3a, and b below, while the SLA process can only print polymer composites with functional additives such as short fibers, as shown in Figure 8.3c. Currently, continuous fibers are blended with thermoplastics or thermoset precursors before the manufacturing process. FDM employs a thermoplastic filament to feed into the heated printing nozzle and extrude the molten material onto a deposition bed with a pre-programmed printing path in a layer-by-layer manner. Continuous fiber is extruded from another nozzle to deposit the fiber on the matrix material. Various composite filaments and FDM printers (e.g., Markforged, Watertown, MA) have been commercialized to print composites with a predetermined resin-fiber combination. Different from FDM, DIW deposits the continuous fiber coated with a reactive thermoset resin on a platform [34–36], and then it will be immediately cured by thermal [34, 37–39] or UV [33, 40, 41] sources to form structures.

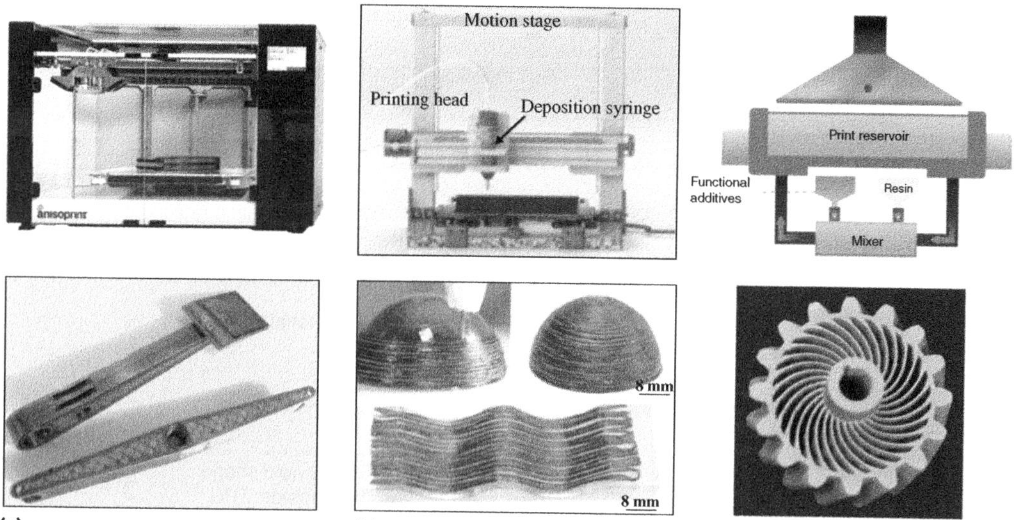

(a) (b) (c)

Figure 8.3 Manufacturing processes of AM-CFRPs: (a) FDM, (b) DIW, and (c) DLP. *Source*: (a) Anisoprint/https://anisoprint.com//last accessed on Oct 10, 2025; (b) [29]/with permission of Elsevier; (c) 3D PRINTING WITH FUNCTIONAL ADDITIVES/3DFortify/https://3dfortify.com/wp-content/uploads/2020/06/Fortify_CKM-Technology_RevA.pdf/.

Continuous Composite Inc. (Coeur d'Alene, ID) developed the DIW 3D printer [32] for photo-curable epoxy composites. The printer head comprises a syringe for resin storage, a deposition nozzle, a feeder for continuous fiber, and an UV lamp for photopolymerization. The feeder connects to the syringe through a one-way valve, which prevents the liquid resin from entering the feeder. In the meantime, the one-way valve provides considerable friction to the fiber extrusion. Therefore, the deposited composite filaments are required to be fully cured upon UV irradiation before the printer head advances.

Fiber-reinforced polymer AM finds applications across various industries. In aerospace, it can produce lightweight, high-strength components for aircraft and spacecraft. In automotive manufacturing, it can create strong yet lightweight parts for vehicles. Infrastructure sectors benefit from their ability to fabricate durable and corrosion-resistant structural elements. Despite its advantages, fiber-reinforced polymer AM presents challenges, including the need for precise fiber alignment, prevention of voids or defects, and post-processing steps like curing and finishing. Research continues to address these challenges and refine the process to unlock the full potential of fiber-reinforced polymer AM.

8.3 Multi-materials Design

The multi-materials design for AM has four steps: (1) the initial design to understand the functional requirement; (2) the process planning design to determine the materials and manufacturing processes; (3) detailed design to optimize the geometry and the material composition; (4) design validation to test the actual performance of the final design.

The design for multi-material additive manufacturing (DfMAM) framework consists of four interconnected modules [42], illustrated in Figure 8.4. The design process begins with identifying functional and technical requirements based on customer needs (Module 1). In the "primary

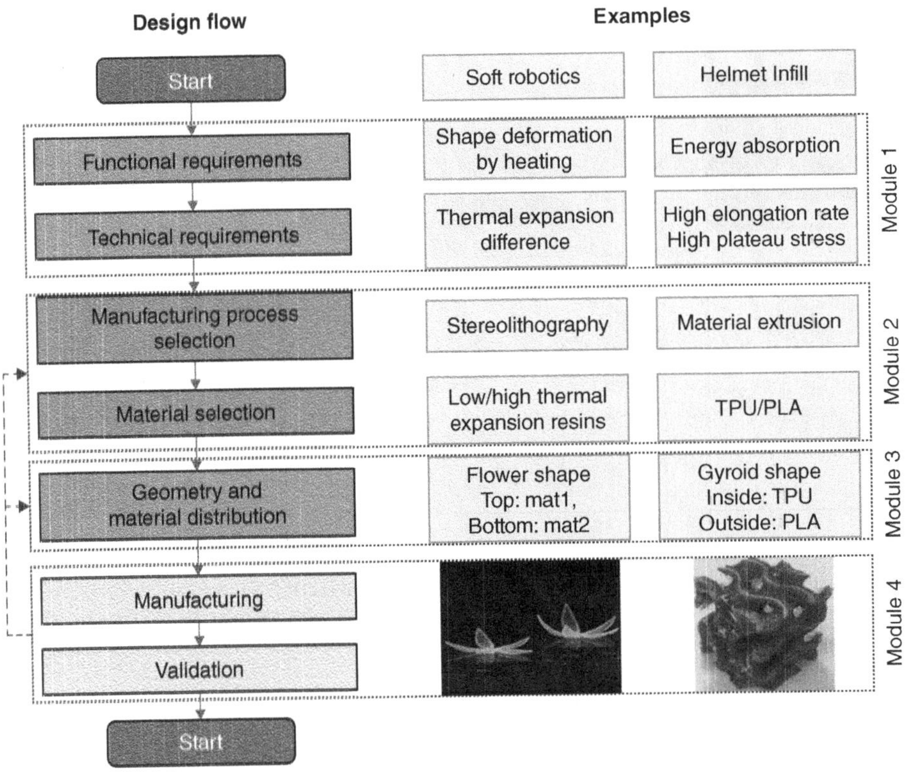

Figure 8.4 The design workflow for multi-material structures. *Source*: Adapted from https://pubs.rsc.org/en/content/articlelanding/2018/sm/c8sm00990b; https://www.sciencedirect.com/science/article/abs/pii/S0020740323006367; [61]/with permission of The Royal Society of Chemistry; [62]/with permission of Elsevier.

material selection" (Module 2), designers choose candidate primary materials (Module 3) from AM material databases to meet multiple design requirements. Appropriate AM processes are then selected to fabricate the product with the chosen materials, and corresponding DfAM rules are derived from the process database. Finally, material composition and part geometry are designed to fulfill functional requirements while adhering to DfAM rules to ensure manufacturability. The dashed lines in Figure 8.4 indicate iterations in primary material and MMAM process selection when the detailed product design (in Module 4) does not yield satisfactory results.

Functional requirements should be identified at the beginning of the DfMAM process. For example, if we are going to design a soft robotic that can change its shape with a thermal actuation, shape deformation by heating is the functional requirement. If the customer requires a helmet that has specified energy absorption properties, the energy absorption rate is a functional requirement. After a functional analysis, we can determine more specific technical requirements. Multiple materials with different thermal expansion rates can be used to design the soft robotic. The specific thermal expansion rate is the most important technique requirement. Materials with high plateau stress and elongation rate can achieve different energy absorption properties; these two properties can be the specific technical requirements for the helmet design. Because these technical requirements can be complex, multi-material parts can have more design freedoms to achieve those requirements than single-material parts.

Once the technical requirements are generated in Module 1, it is time to choose the appropriate manufacturing process and materials in the primary material selection module. The manufacturing process will be chosen first because only when the process is determined, the spectrum of available materials for that process can be selected. For example, the soft robotics has a small size but requires high resolution. SLA is selected for manufacturing soft robotics. However, the helmet has a relatively large size, and if the client would like to lower the cost, FDM is selected as the manufacturing process. To enable the shape-changing of soft robotics, high and low thermal expansion resins can be chosen to print the multi-material soft robotics. The materials to make the helmet require a high plateau stress and a large elongation rate. TPU can achieve these properties, but the stress in TPU is low, which limits the energy absorption capability. PLA is a rigid material that can increase the plateau stress of a multi-material structure. Therefore, PLA and TPU are selected for energy-absorption helmet printing. However, extra constraints specific to multi-material part design must be taken into account during primary material selection. One critical consideration is the mutual compatibility of dissimilar materials that need to be bonded to create an integrated part. Only materials that form a well-bonded interface can be considered suitable primary materials. Therefore, mutual compatibility testing acts as a filter in the material selection process. An AM-specific material database is essential for material selection. This database should be organized by material compatibilities, as each AM technique can process only a limited range of material families. In some cases, using one material in the product might negatively impact the performance of other primary materials. This trade-off effect must be considered when selecting materials for multi-material parts. In both examples, the interfacial strength between two different materials should be tested to ensure the integrity of the multi-material structure.

After the selection of the manufacturing process and material candidates, we need to design the geometry and material distribution of the structure. First, the model can be divided into zones based on different functional requirements. Each zone is assigned materials that best suit its specific needs. For instance, areas that must withstand high stress should be made from materials with high tensile strength, while flexible regions should use materials with high elasticity. Then, simulations, such as finite element analysis (FEA), can be used to assess structural performance under load conditions and material behavior simulations to ensure compatibility. Optimize the geometry and material distribution through iterative topology and material optimization processes. For the soft robotic design, if the structure needs to bend, we can put materials with a high thermal expansion rate at lower levels of the structures and put materials with a low thermal expansion rate at top levels. For the helmet design, we can adopt gyroid lattice structures with PLA on the outer walls and TPU on the inner walls to achieve high energy absorption capabilities.

After finalizing the design and material distribution, we can proceed to manufacturing and validation. First, finalize the CAD models and ensure compatibility with the 3D printer. Select the appropriate multi-material 3D printer and slice the CAD model to generate the machine code. Prepare and calibrate the printer for each material, ensuring proper settings for nozzle and bed temperatures in the FDM process or the light intensity and exposure time in the SLA process. During the manufacturing process, periodically inspect the print for issues, as multi-material parts have a higher failure rate than single-material parts. Validation of the multi-material structure is an important step that can verify the actual performance of the structure compared to the functional requirement. For example, the bending angle of soft robotics can be measured at a certain temperature to quantify the shape-changing capability. The stress-strain curve of the gyroid structure can be obtained via a compression mechanical test. The energy absorption rate can be calculated from the stress-strain curve. If a large discrepancy is found between the experimental validation and the simulation models used in module 3, the simulation model should be reexamined to adjust the geometry and material distribution in the structure. If the problem still exists, we should go back further to the process and material selection to check if other options can achieve better results. Therefore, the DfMAM process is an iterative process that requires multiple loops to improve the performance of multi-material structures.

8.4 Voxel Printing Design

Voxel printing in AM involves creating objects layer by layer at the voxel level, where each voxel represents a tiny volume of the object, often just a few micrometers on a side. This method allows for highly detailed control over the material properties and structure. The design methodology for voxel printing requires a 3D density map in general. In the literature, medical scanning data has been widely used in the design process of voxel-printed objects. For example, precise voxel-level manipulation in multi-material 3D printing can enable the seamless integration of diverse medical data types like functional magnetic resonance imaging, tractography, and four-dimensional flow onto conventional magnetic resonance imaging and computed tomography datasets [43]. An example of using MRI data to do a voxel printing is shown in Figure 8.5. The MRI data is a series of images with values in each pixel, which can be represented by grayscale images, as shown in Figure 8.5a. These data not only capture the anatomy but also reflect the velocity of blood flow. A data thresholding technique is used to screen out irrelevant data and keep meaningful data to define the solid region of the blood pool, as shown in Figure 8.5b, and c. The value of the data in the blood pool means the velocity of the blood, which is mapped to the color of the material. Each vertex is assigned to a color as shown in Figure 8.5d, and the voxel color is then determined from its vertex, as shown in Figure 8.5e. The color map is then added to the segmentation to construct the composite data, as shown in Figure 8.5f. The final voxel printing model is shown in Figure 8.5g, which shows the flow velocity in the cardiac.

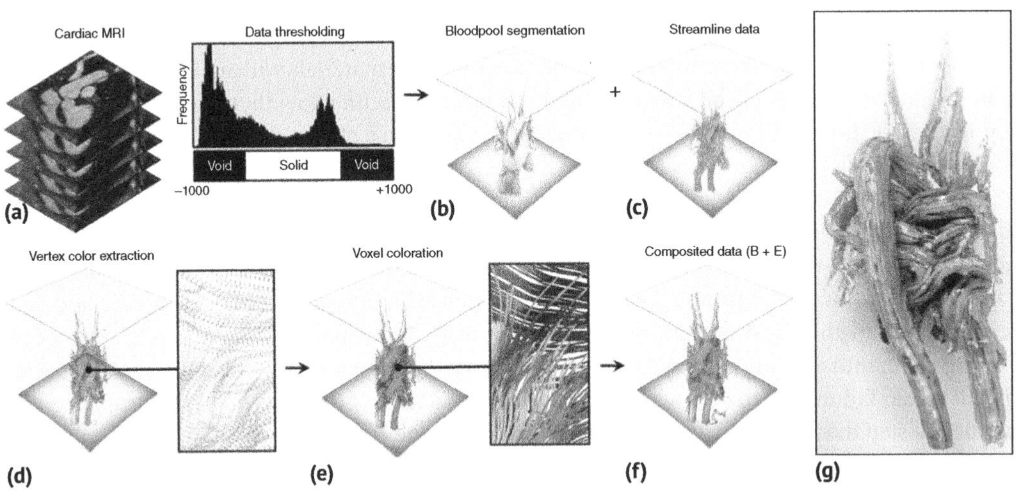

Figure 8.5 Voxel printing by MRI images. *Source*: [43]/Sage Publications/CC BY 4.0 https://www.liebertpub.com/doi/full/10.1089/3dp.2022.0265.

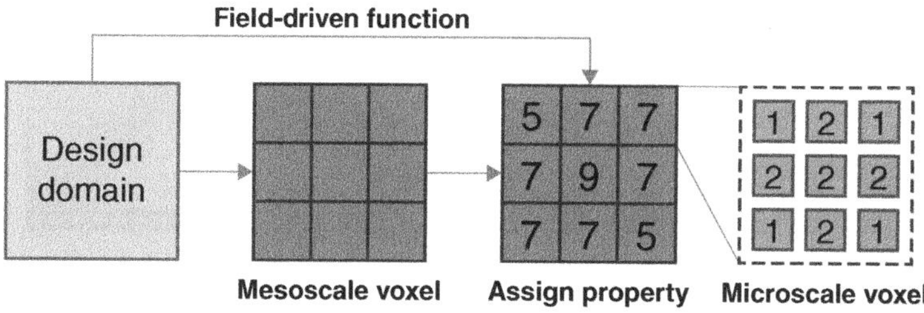

Figure 8.6 The design workflow of two-scale field-driven design methods for voxel printing.

However, if there is no direct data like the 4D MRI, it is very complicated to design every voxel in the model from scratch. That is why voxel printing design methods are very limited. In general, if we want to design a voxel-printed part, we need to discretize the geometry to voxels with a dimension around 20 μm. A small part can contain millions of voxels that challenge the design process. Two-scale field-driven design methods can be used to simplify the design process. The workflow of this design method is shown in Figure 8.6. Based on the functional requirement, an objective function can be established by applying a field variable in the design domain. The design domain is then discretized to mesoscale voxels. The mesoscale voxel contains multiple microscale voxels. Because voxel printing can only assign limited types of materials to each voxel. This mesoscale voxel can extend the material property options by changing the combinations of microscale voxels. The value of each mesoscale voxel can be calculated by taking the coordinates of the center into the field function. This value can be one or multiple material properties. The second step is to find a microscale voxel combination that can achieve the desired material properties in the mesoscale voxel. The size of the microscale voxel is the same as the voxel size of the printer. In the microscale voxel, the integers represent the type of material used in that voxel. The information in the microscale voxel can be directly sent to the printer to start the voxel printing. However, to implement this workflow, a material database is needed that can map the material property to the combination of microscale voxels. Machine learning is a promising tool to establish this relationship.

8.5 Fiber-reinforced Composite Design

The flexibility of the fiber placement provided by AM processes brings promising opportunities for mechanical performance optimization. Unlike traditional composite manufacturing techniques, fiber-reinforced composites made by AM can achieve unprecedented mechanical properties by optimizing the fiber orientation and volume fraction. Continuous fiber orientation optimization approaches have been studied for two decades [44–46]. Recently, approaches to optimize the fiber trajectory in composites have been developed, such as fluid streamlines [47, 48], maximum principal stress [49–52], and pointing stress vectors [53, 54]. These approaches assigned all fibers within composites with a single orientation and identical fiber volume fraction. The topology optimization method has also been used to optimize the fiber and the geometry simultaneously [55–58]. The constitutive matrix of the finite element is derived by the volume fraction and the fiber orientation [59], which are optimized simultaneously as shown in Figure 8.7.

Figure 8.7 Optimization of the fiber orientation and the topology of AM-CCFRPs. *Source*: [59]/MDPI/CC BY 4.0.

The maximum principal stress is one of the most widely used design methods for continuous fiber-reinforced composites made by AM. The first step is to do a FEA for the initial design model. The stress tensor in each element can be obtained from the FEA result. The principal stress direction can be calculated from the stress tensor. Because the continuous fiber is placed layer by layer in the AM process, the 2D stress tensor and principal stress direction are enough for fiber path design. However, it should be noted that some processes, such as AFP, can place fibers on curved surfaces. 3D stress tensor and principal stress direction might be needed for a more complicated design. Once the principal stress direction in each element is found, a vector field can be generated based on the principal stress directions. This vector field can be used to generate streamlines that can guide the placement of continuous fibers.

Although the maximum principal stress method can design the placement of fibers, it may not lead to an optimal design. Because the principal stresses are obtained in the FEA model of the original design. The principal stresses are only valid in that circumstance. Once the fiber is added to the part, the principal stress will change. Therefore, an iterative design process is needed to optimize the placement of fibers. The orientation optimization method can be used to find the optimal orientation of each element that can be used to design the fiber path. In this method, each element is assigned a density variable and an orientation angle variable. The constitutive matrix of the element is determined by the density and the orientation of the element. A topology optimization algorithm can be used to optimize the density and orientation concurrently. When the design variables are converged, the final orientations can be used to generate a vector field and streamlines, which can be used to define the fiber placement in the printing process.

8.6 Summary

Digital materials, which can be digitally controlled and manufactured, contrast with traditional homogeneous materials like plastics and alloys by enabling complex and multifunctional designs, such as soft robotics that require functionally graded materials for shape transformation. AM accelerates the development of these digital materials through multi-material printing, voxel printing, and fiber-reinforced polymer printing. Multi-material AM techniques like material jetting, extrusion, SLA, and PBF allow for the creation of intricate structures with tailored properties, enhancing mechanical performance and integrating multiple functionalities. Voxel printing offers precise control over material deposition at the voxel level, resulting in highly detailed and customized objects. Fiber-reinforced polymer AM, using both continuous and short fibers, produces lightweight, high-strength components for industries like aerospace and automotive. The design process for multi-material structures involves defining functional requirements, selecting appropriate materials and processes, optimizing geometry and material distribution, and validating the final product. Voxel printing design utilizes 3D density maps, often from medical scans, for precise and functional structures, leveraging advanced design methodologies and machine learning to manage complexity. The fiber-reinforced composite design focuses on optimizing fiber placement and orientation, using techniques like FEA and topology optimization to align fiber paths with principal stress directions for optimal strength. This chapter highlights the advancements and methodologies in digital materials that leverage the material complexity enabled by AM, emphasizing their potential for innovative applications and enhanced performance across various industries. In Chapter 9, the application of AI and machine learning in DfAM will be explored. Data-driven design approaches with advanced AI models have the potential to revolutionize DfAM in the next decade.

References

1 Ge, Q., et al., *Multimaterial 4D printing with tailorable shape memory polymers*. Scientific Reports, 2016. **6**(1): pp. 1–11.

2 Bandyopadhyay, A. and Heer, B., *Additive manufacturing of multi-material structures*. Materials Science and Engineering: R: Reports, 2018. **129**: pp. 1–16.

3 Nazir, A., et al., *Multi-material additive manufacturing: a systematic review of design, properties, applications, challenges, and 3D printing of materials and cellular metamaterials*. Materials & Design, 2023. **226**: p. 111661.

4 Li, Y., et al., *A review on functionally graded materials and structures via additive manufacturing: from multi-scale design to versatile functional properties*. Advanced Materials Technologies, 2020. **5**(6): p. 1900981.

5 Zhang, M., et al., *Mechanical properties of multi-materials porous structures based on triply periodic minimal surface fabricated by additive manufacturing*. Rapid Prototyping Journal, 2021. **27**(9): pp. 1681–1692.

6 Singh, R., et al., *Multi-material additive manufacturing of sustainable innovative materials and structures*. Polymers, 2019. **11**(1): p. 62.

7 Hasanov, S., et al., *Review on additive manufacturing of multi-material parts: progress and challenges*. Journal of Manufacturing and Materials Processing, 2022. **6**(1): p. 4.

8 Vidimce, K., et al., Foundry: hierarchical material design for multi-material fabrication. in *Proceedings of the 29th Annual Symposium on User Interface Software and Technology*, 2016, Association for Computing Machinery: Tokyo, Japan. pp. 563–574.

9 Elkaseer, A., et al., *Material jetting for advanced applications: a state-of-the-art review, gaps and future directions*. Additive Manufacturing, 2022. **60**: p. 103270.

10 Boulaala, M., et al., *Towards design of mechanical part and electronic control of multi-material/multicolor fused deposition modeling 3D printing*. The International Journal of Advanced Manufacturing Technology, 2020. **110**(1): pp. 45–55.

11 Saadi, M.A.S.R., et al., *Direct ink writing: a 3D printing technology for diverse materials*. Advanced Materials, 2022. **34**(28): p. 2108855.

12 Wicker, R.B. and MacDonald, E.W., *Multi-material, multi-technology stereolithography*. Virtual and Physical Prototyping, 2012. **7**(3): pp. 181–194.

13 Schneck, M., et al., *Capability of multi-material laser-based powder bed fusion–development and analysis of a prototype large bore engine component*. Metals, 2022. **12**(1): p. 44.

14 Cheng, J., et al., *Centrifugal multimaterial 3D printing of multifunctional heterogeneous objects*. Nature Communications, 2022. **13**(1): p. 7931.

15 Chadha, C., et al. Design for multi-material manufacturing using polyjet printing process: a review. in *ASME 2022 International Design Engineering Technical Conferences and Computers and Information in Engineering Conference*. 2022.

16 van de Werken, N., et al., *Additively manufactured carbon fiber-reinforced composites: state of the art and perspective*. Additive Manufacturing, 2020. **31**: p. 100962.

17 Eckel, Z.C., et al., *3D printing additive manufacturing of polymer-derived ceramics*. Science, 2016. **351**(6268): pp. 58–62.

18 Zarek, M., et al., *3D printing of shape memory polymers for flexible electronic devices*. Advanced Materials, 2016. **28**(22): pp. 4449–4454.

19 Tumbleston, J.R., et al., *Continuous liquid interface production of 3D objects*. Science, 2015. **347**(6228): pp. 1349–1352.

20 Ge, Q., Qi, H.J., and Dunn, M.L., *Active materials by four-dimension printing*. Applied Physics Letters, 2013. **103**(13): p. 131901.

21 Mao, Y.Q., et al., *Sequential self-folding structures by 3D printed digital shape memory polymers*. Scientific Reports, 2015. **5**: p. 13616.

22 Fernandes, R.R., et al., *Experimental investigation of additively manufactured continuous fiber reinforced composite parts with optimized topology and fiber paths*. Additive Manufacturing, 2021. **44**: p. 102056.

23 Pappas, J.M., et al., *A comparative study of pellet-based extrusion deposition of short, long, and continuous carbon fiber-reinforced polymer composites for large-scale additive manufacturing*. Journal of Manufacturing Science and Engineering, 2021. **143**(7): p. 071012.

24 Li, N., Link, G., and Jelonnek, J., *3D microwave printing temperature control of continuous carbon fiber reinforced composites*. Composites Science and Technology, 2020. **187**: p. 107939.

25 Matsuzaki, R., et al., *Three-dimensional printing of continuous-fiber composites by in-nozzle impregnation*. Scientific Reports, 2016. **6**(1): p. 23058.

26 Liu, T., et al., *Interfacial performance and fracture patterns of 3D printed continuous carbon fiber with sizing reinforced PA6 composites*. Composites Part A: Applied Science and Manufacturing, 2018. **114**: pp. 368–376.

27 Liu, T., et al., *High-pressure interfacial impregnation by micro-screw in-situ extrusion for 3D printed continuous carbon fiber reinforced nylon composites*. Composites Part A: Applied Science and Manufacturing, 2020. **130**: p. 105770.

28 Thakur, A. and Dong, X., *Printing with 3D continuous carbon fiber multifunctional composites via UV-assisted coextrusion deposition*. Manufacturing Letters, 2020. **24**: pp. 1–5.

29 He, X., et al., *3D printing of continuous fiber-reinforced thermoset composites*. Additive Manufacturing, 2021. **40**: p. 101921.

30 Nardiello, J.A., et al., *Device and method for 3D printing with long-fiber reinforcement*. 2019, Google Patents. US10173410B2.

31 Lewicki, J., et al., *Additive manufacturing continuous filament carbon fiber epoxy composites*. 2017, Google Patents. US20170015060A1.

32 Tyler, K., *Method and apparatus for continuous composite three-dimensional printing*. 2016, Google Patents. US20140061974A1.

33 Rahman, M.A., et al., *3D printing of continuous carbon fiber reinforced thermoset composites using UV curable resin*. Polymer Composites, 2021. **42**(11): pp. 5859–5868.

34 Lewis, J.A., *Direct ink writing of 3D functional materials*. Advanced Functional Materials, 2006. **16**(17): pp. 2193–2204.

35 Lewis, J.A. and Gratson, G.M., *Direct writing in three dimensions*. Materials Today, 2004. **7**(7-8): pp. 32–39.

36 Gratson, G.M., Xu, M.J., and Lewis, J.A., *Microperiodic structures - direct writing of three-dimensional webs*. Nature, 2004. **428**(6981): pp. 386–386.

37 Compton, B.G. and Lewis, J.A., *3D-printing of lightweight cellular composites*. Advanced Materials, 2014. **26**(34): pp. 5930–5935.

38 Gladman, A.S., et al., *Biomimetic 4D printing*. Nature Materials, 2016. **15**(4): pp. 413–418.

39 Lewis, J.A., Novel inks for direct writing in three dimensions. in *Digital Fabrication 2005, Final Program and Proceedings*, 2005. pp. 142–142.

40 Ambulo, C.P., et al., *Four-dimensional printing of liquid crystal elastomers*. ACS Applied Materials & Interfaces, 2017. **9**(42): pp. 37332–37339.

41 Kotikian, A., et al., *3D printing of liquid crystal elastomeric actuators with spatially programed nematic order*. Advanced Materials, 2018. **30**: p. 1706164.

42 Yao, X., et al., *A multi-material part design framework in additive manufacturing*. The International Journal of Advanced Manufacturing Technology, 2018. **99**: p. 2111–2119.

43 Jacobson, N.M., et al., *The inner complexities of multimodal medical data: bitmap-based 3D printing for surgical planning using dynamic physiology*. 3D Printing and Additive Manufacturing, 2023. **10**(5): pp. 855–868.

44 Cheng, H.C., Kikuchi, N., and Ma, Z.D., *An improved approach for determining the optimal orientation of orthotropic material*. Structural Optimization, 1994. **8**(2): pp. 101–112.

45 Pedersen, P., *On optimal orientation of orthotropic materials*. Structural Optimization, 1989. **1**(2): pp. 101–106.

46 Luo, J.H. and Gea, H.C., *Optimal orientation of orthotropic materials using an energy based method*. Structural Optimization, 1998. **15**(3): pp. 230–236.

47 Yamanaka, Y., et al., *Fiber line optimization in single ply for 3D printed composites*. Open Journal of Composite Materials, 2016. **6**(04): p. 121.

48 Khan, S., et al., *Curvilinear variable stiffness 3D printing technology for improved open-hole tensile strength*. Additive Manufacturing, 2018. **24**: pp. 378–385.

49 Sugiyama, K., et al., *3D printing of optimized composites with variable fiber volume fraction and stiffness using continuous fiber*. Composites Science and Technology, 2020. **186**: p. 107905.

50 Zhang, H., Yang, D., and Y. Sheng, *Performance-driven 3D printing of continuous curved carbon fibre reinforced polymer composites: a preliminary numerical study*. Composites Part B: Engineering, 2018. **151**: pp. 256–264.

51 Hou, Z., et al., *Optimization design and 3D printing of curvilinear fiber reinforced variable stiffness composites.* Composites Science and Technology, 2021. **201**: p. 108502.

52 Malakhov, A.V., et al., *A modeling method of continuous fiber paths for additive manufacturing (3D printing) of variable stiffness composite structures.* Applied Composite Materials, 2020. **27**(3): pp. 185–208.

53 Suzuki, T., Fukushige, S., and Tsunori, M., *Load path visualization and fiber trajectory optimization for additive manufacturing of composites.* Additive Manufacturing, 2020. **31**: p. 100942.

54 Zhao, S., Wu, N., and Wang, Q., *Load path-guided fiber trajectory in composite panels: a comparative study and a novel combined method.* Composite Structures, 2021. **263**: p. 113689.

55 Schmidt, M.-P., et al., *Structural topology optimization with smoothly varying fiber orientations.* Structural and Multidisciplinary Optimization, 2020. **62**(6): pp. 3105–3126.

56 Boddeti, N., et al., *Optimal design and manufacture of variable stiffness laminated continuous fiber reinforced composites.* Scientific Reports, 2020. **10**(1): p. 16507.

57 Desai, A., et al., *Topological-derivative-based design of stiff fiber-reinforced structures with optimally oriented continuous fibers.* Structural and Multidisciplinary Optimization, 2021. **63**(2): pp. 703–720.

58 Kim, D., et al., *Topology optimization of functionally graded anisotropic composite structures using homogenization design method.* Computer Methods in Applied Mechanics and Engineering, 2020. **369**: p. 113220.

59 Jiang, D., Hoglund, R., and Smith, D.E., *Continuous fiber angle topology optimization for polymer composite deposition additive manufacturing applications.* Fibers, 2019. **7**(2): p. 14.

60 Christoph Bader et al., *Making data matter: Voxel printing for the digital fabrication of data across scales and domains.* Sci. Adv. 4, eaas8652 (2018). https://doi.org/10.1126/sciadv.aas8652.

61 van Rees, Wim M., et al. *Mechanics of Biomimetic 4D Printed Structures.* Soft Matter, vol. 14, no. 43, 2018, pp. 8771–79.

62 Teawdeswan, Ladpha, and Guoying Dong. I*nverse Design of Multi-material Gyroid Structures Made by Additive Manufacturing.* International Journal of Mechanical Sciences, vol. 262, 2024, p. 108734.

9

Data-driven and Machine Learning-based Design for Additive Manufacturing

Additive manufacturing (AM) offers unprecedented design freedom—enabling complex geometries, functional integration, and material efficiency. However, this very freedom challenges conventional design practices and creates a critical demand for new tools and methodologies. As discussed in Chapter 1, design for additive manufacturing (DfAM)—as defined in this book—is the integrative process of creating, evaluating, and optimizing product geometry, structure, and material composition by explicitly accounting for the unique capabilities and limitations of additive manufacturing processes. DfAM includes two complementary dimensions:

- Restrictive DfAM, which ensures manufacturability through rules and constraints tied to specific AM technologies, materials, and post-processing requirements;
- Opportunistic DfAM, which explores novel functional, geometric, and material possibilities uniquely enabled by AM, such as lattice structures, part consolidation, and functionally graded materials.

Increasingly, DfAM is becoming data-driven and computational, relying on simulation, performance prediction, and machine learning (ML) to guide design decisions, automate screening, and generate or optimize designs that balance performance, cost, and manufacturability [1].

ML is not a replacement for design expertise but rather it is an amplifier of it. ML methods can identify patterns across large and complex design spaces, predict print and performance outcomes, accelerate computational workflows such as topology optimization (TO), and enable adaptive decision support during early-stage design [2]. These capabilities support both restrictive and opportunistic DfAM strategies: from manufacturability assessment and constraint-aware optimization to functional discovery and generative design of architected materials. ML is thus emerging as a transdisciplinary enabler across the DfAM-related domains such as design, process, and materials.

Despite its promise, the integration of ML into DfAM presents new challenges. Data scarcity, model generalizability, and interpretability currently limit deployment in real-world industrial settings. Moreover, many ML applications remain narrowly scoped, focusing on isolated tasks, such as design candidate classification and design concept generation, without integration into holistic design workflows. To overcome these limitations, the next generation of DfAM must embrace hybrid approaches [2, 3] that combine ML with various other potential approaches such as physics-based modeling, emphasize explainability and robustness, and build interoperable digital infrastructures to support data sharing and reuse.

This chapter presents a thorough overview of data-driven and ML-based approaches in DfAM, with a focus on how ML is reshaping the ways designs are conceptualized, evaluated, and optimized for AM. Rather than treating ML as a standalone technique, we frame it as an intelligence tool embedded within and across DfAM strategies. Drawing on current research and industrial applications, we classify ML contributions into four interrelated areas introduced in Chapter 1:

- Topology optimization and performance-driven design.
- Conceptual design and generative modeling.
- Manufacturability and printability prediction.
- Process-aware design and property estimation.

Through this structure, the chapter explores how ML supports the broader evolution of DfAM. The chapter concludes with a discussion of current limitations and future research opportunities, including the development of explainable artificial intelligence (AI), hybrid modeling frameworks, and open data standards to support scalable and trustworthy ML in DfAM.

9.1 Foundations of Data-driven and ML-based DfAM

The unique characteristics of AM, such as its ability to fabricate highly complex geometries, its sensitivity to process parameters, and its tight coupling between material behavior, process conditions, and part performance, create a design space that is difficult to navigate using traditional rule-based or analytical approaches. Unlike conventional manufacturing, where decades of empirical knowledge have led to well-established design heuristics, DfAM often involves the creation of novel structures, the exploration of unconventional design freedoms, and the management of constraints that vary significantly across AM technologies and materials. These challenges make DfAM particularly well suited for ML approaches, which can capture complex, nonlinear relationships and support predictive or generative decision-making based on data [1]. These methods enable the extraction of useful patterns from high-dimensional design spaces, prediction of print outcomes, and generation of manufacturable yet innovative design candidates.

Data-driven design refers to the use of empirical or simulated data to inform or automate design decisions. In the context of DfAM, this data can originate from a variety of sources, including potential relationships between design parameters and performance outcomes, outputs from computational simulations such as finite element analysis (FEA), and detailed records of manufacturing process parameters and print logs. Additionally, material property databases and geometry–performance–property maps—particularly those developed for architected structures such as lattices or metamaterials—serve as rich resources for guiding decision-making. These diverse data streams collectively enable the development of predictive models and design automation tools tailored to the complexities of AM [4, 5]. Unlike purely rule-based or analytical approaches, data-driven methods can generalize across complex input spaces, learn from examples, and adapt to new conditions—provided sufficient and representative data are available.

ML is a subset of AI that focuses on algorithms capable of improving their performance through exposure to data. Within the context of DfAM, ML models are employed for a range of tasks that support both performance evaluation and design automation [3]. These include predicting mechanical or thermal responses from geometric features or process settings, assessing the printability or manufacturability of candidate designs, accelerating computationally intensive tasks such as TO, and enabling design space exploration or automated generation of complex architectures. By serving as surrogate models or intelligent decision aids, ML techniques help reduce dependence on time-consuming simulations and manual design iterations, thereby enhancing the efficiency and responsiveness of the design process [6].

Several categories of ML are relevant to DfAM tasks:

- **Supervised Learning**: This type of models learns a mapping from input features (e.g., geometric descriptors, process parameters) to target outputs (e.g., strength, porosity, manufacturability). Typical algorithms include neural networks, decision trees, support vector machines, etc. [7]. For example, supervised learning models such as Gaussian processes and classification and regression trees (CART) have been used to construct process-aware design rules or predict turbulent flow behaviors in functionally optimized parts like heat exchangers [8, 9].
- **Unsupervised Learning**: This type of models extracts patterns or clusters from unlabeled data. Often used to explore geometry or material property spaces, or to group designs based on similarity. For example, generative models—including conditional Generative Adversarial Networks (cGANs)—have been developed to produce topology-optimized geometries directly, bypassing costly iterative simulations [10].
- **Reinforcement Learning**: These models are useful in generative design or TO settings where design sequences are explored via a reward mechanism (e.g., compliance reduction, support minimization). It supports sequential decision-making under uncertainty [2].
- **Hybrid and Physics-informed ML**: These methods combine ML methods with domain knowledge or simulation-based constraints. For example, ML may predict intermediate TO results while embedding manufacturability filters, or fuse image-based features with process physics [5, 6]. Hybrid ML model combining neural network (NN) and modified convolutional neural network (CNN) [11] can process CAD file and AM process information simultaneously to identify manufacturability issues to be fabricated using a selected AM process.

9.2 ML Across DfAM Workflow

The integration of ML into DfAM is not confined to a single stage of the design process. Rather, it spans a spectrum of tasks—from performance-driven optimization to manufacturability screening—providing computational support for both restrictive and opportunistic strategies. Each approach offers different strengths depending on the availability of data, the complexity of the task, and the degree of interpretability or reliability required.

9.2.1 Topology Optimization and Performance-driven Design

TO is one of the most mature areas of DfAM where ML has had demonstrable impact. Traditional TO methods are computationally intensive and often require hundreds of iterations to converge to a high-performance design [10]. TO is typically used to create lightweight and high-performance designs as a part of the opportunistic strategy. ML techniques—especially deep neural networks (DNNs), convolutional neural networks (CNNs), and generative adversarial networks (GANs)—are increasingly used to accelerate or bypass these iterations by learning mappings from boundary conditions and constraints to optimized geometries. Several works have shown the potential of ML to approximate intermediate TO results, generate design candidates under multi-objective constraints, or perform inverse mapping from desired performance to geometry. Reinforcement learning (RL) methods, such as those using upper confidence bound (UCB) strategies, have also been introduced to explore the TO design space while incorporating additive manufacturing constraints like support minimization and thin feature suppression [2]. A study on the design of liquid-to-air heat exchangers using a density-based TO algorithm has demonstrated the potential of ML in DfAM [8]. The study showed that the integration of ML-based turbulent flow modeling and DfAM constraints can lead to significant improvements in power density [8].

9.2.2 Conceptual Design and Generative Modeling

In early-stage design, ML is increasingly used to automate ideation, explore unconventional architectures, and embed domain knowledge into the conceptualization process. This includes the generation of functionally graded lattice structures, architected materials, and bio-inspired geometries. Generative models—such as variational autoencoders (VAEs), conditional GANs (cGANs), and transformer-based networks—have been applied to produce novel geometries that are structurally valid and optimized for performance or compliance. Integration with knowledge graphs has also been proposed to capture reusable DfAM knowledge and apply it during the conceptual design process [12]. We also developed hybrid CVAE-DDPM models for generating auxetic metamaterials, allowing one-shot generation of topologically valid, performance-optimized designs [13]. Such generative models often are combined with design optimization to further improve the performance of the generated design as shown in Figure 9.1. These methods exemplify how ML expands the opportunistic DfAM space by enabling rapid exploration of nonintuitive, multifunctional geometries that traditional heuristics may overlook.

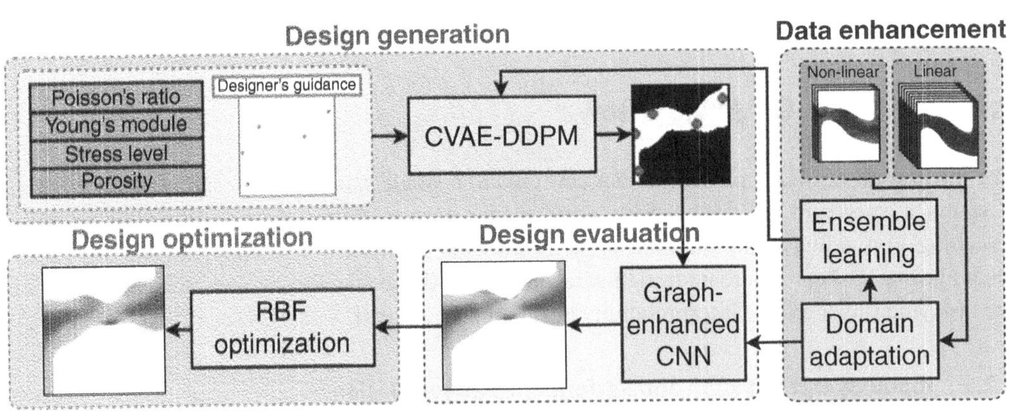

Figure 9.1 Workflow of the generative and surrogate models for auxetic metamaterials design and optimization. *Source*: [13]/with permission of Elsevier.

9.2.3 Manufacturability and Printability Prediction

From the restrictive DfAM perspective, ML enables the formalization of manufacturability constraints, such as minimum feature sizes or overhang angles, wall thickness limits, and support structure requirements, which can be integrated into TO frameworks [14]. Additionally, these manufacturing constraints can be used to drive standalone feasibility evaluations. Expert systems and statistical rule-learning approaches also play a key role in automated design verification, classifying designs as compliant or noncompliant with established AM constraints using interpretable models such as decision trees and SVMs [15]. Automated design verification is another important aspect of manufacturability and printability analysis. A DfAM expert system that evaluates part compliance with established AM design rules has been developed that uses a decision tree algorithm to classify designs as either compliant or noncompliant [16]

Our lab developed a decision-support system that uses CAD-extracted features to assess AM candidacy at both part and assembly levels [17]. The system employs supervised ML models to classify parts based on geometric complexity, support needs, and cost trade-offs, delivering actionable feedback for designers with limited AM experience. This approach addresses key gaps in early manufacturability assessment and promotes restrictive DfAM compliance across diverse design settings. A web-based application of the trained ML models was implemented with its main interface shown in Figure 9.2 following the workflow of (1) load CAD model, (2) new data acquisition, and (3) candidacy prediction. The developed web application is able to load common CAD files and automatically extract geometric information of volume, surface area, and number of components. Then, users are requested to fill missing data of model information (e.g., number of interfaces), model features (e.g., whether exist internal channels), and economic considerations (e.g., lead time). With these inputs, the tool produces candidacy scores for each potential (see Figure 9.2c). To learn about the specific AM potential, tutorials can be accessed by hovering above the item (e.g., lightweight bar in Figure 9.2d). These tutorials will help the novice AM users to develop a better understanding of the potential gains by using AM.

9.2.4 Process-aware Design and Property Estimation

ML is also increasingly used to predict process outcomes and material responses, closing the loop between design, process parameter selection, and part property prediction. This is particularly relevant for AM processes characterized by strong dependencies between design geometry, scan strategies, thermal behavior, and structural integrity. Surrogate models, such as Gaussian processes and graph-based CNNs, have been developed to estimate stress distributions, porosity, and other key metrics from combined design and process parameters [3]. As seen in Figure 9.1, property estimation is often part of the generative optimization loop to improve the performance of the design generated for AM. Specifically for auxetic structures, we developed a multi-stage ML framework for predicting mechanical properties of auxetic structures incorporating 2D-to-3D mapping, graph-informed CNNs, and mixture density networks (MDNs) [13, 18]. The framework combines:

- Graph-informed convolutional neural networks (G-CNNs) to extract stress-relevant topological features from 2D unit cell layouts.
- MDNs to predict the probability distribution of stress fields based on both geometric and physical descriptors.

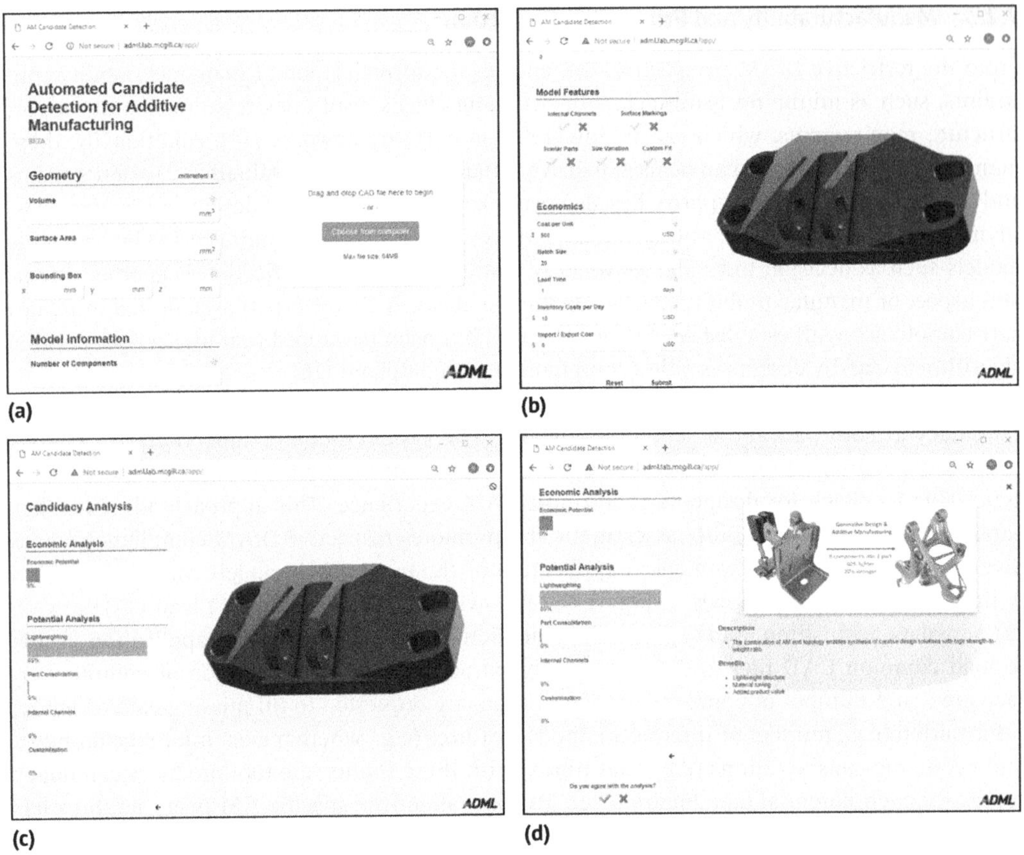

Figure 9.2 Interfaces of the implemented automated candidacy detection for AM (ACDAM) web application. (a) Load CAD model, (b) data acquisition, and (c) candidacy prediction, and (d) Lightweighting potential tutorials.

- A 2D-to-3D geometric conversion pipeline, using open-source CAD tools, that translates 2D auxetic patterns into fully printable 3D structures while preserving stress-optimized design intent.
- A surrogate-based stress optimization loop, allowing rapid screening and refinement of design candidates without resorting to full finite element simulations.

This framework enables efficient design–process–performance coupling for architected materials. It supports inverse design (i.e., identifying input geometry that achieves a target stress behavior) and reduces reliance on extensive physical or simulation data by learning topological priors. It also bridges early-stage lattice conceptualization with downstream manufacturable realization.

9.3 Cross-cutting Strategies and Future Directions

While ML has demonstrated strong potential across design stages in additive manufacturing, its widespread integration into DfAM practice faces several persistent challenges. These are not isolated to any one application area but reflect structural barriers and methodological gaps common across TO, generative design, manufacturability prediction, and property estimation. This section outlines three research directions seeing rising in DfAM field—hybrid ML development, explainable ML development, and data and digital infrastructure development—that are crucial for realizing the next stage of ML-enabled DfAM.

A central challenge in current ML applications is generalizability: models trained on narrow datasets or synthetic geometries often fail to extrapolate to new shapes, materials, or AM processes. To address this, there is growing momentum toward hybrid models that integrate ML with domain-specific knowledge, including:

- Physics-based constraints (e.g., equilibrium, conservation laws) embedded into neural network architectures.
- Rule-informed learning frameworks that encode DfAM constraints (e.g., overhang limits, support minimization) as loss functions or post-processing filters.
- Multi-fidelity learning, where coarse simulation outputs are used to pre-train models before fine-tuning with higher-fidelity or experimental data.

Another key limitation of current ML systems in DfAM is their opacity. Most deep learning models act as "black boxes," making it difficult to understand or verify why certain design decisions are made. This makes application in safety-critical domains such as aerospace and medical devices very difficult to adopt ML-enabled approaches. Recent research has emphasized the development of explainable ML (XAI) techniques tailored for engineering design. These include:

- Feature attribution methods (e.g., SHapley Additive exPlanations (SHAP), Grad-CAM) for visualizing model attention on geometric regions.
- Surrogate-assisted interpretation tools (e.g., Local Interpretable Model-agnostic Explanations (LIME), symbolic regression) that approximate black-box behavior using interpretable decision trees or symbolic regressors.
- Integration of confidence intervals and uncertainty quantification, particularly for inverse design and property prediction tasks.

Many ML models in DfAM remain siloed, trained on local datasets with limited scope or documentation. The lack of standardized data formats, annotated benchmarks, and open APIs hinders replication, reuse, and cross-domain learning. This issue is compounded by the fragmentation of AM design toolchains, such as the outdated file formats of STL to represent design geometry that strip away semantic metadata. To enable scalable, general-purpose ML-DfAM workflows, the field must move toward:

- Public benchmark datasets, including labeled lattices, TO results, process-parameter-property mappings, and failed printing records.
- Adoption of next-generation file formats (e.g., AMF, 3MF) that retain information about tolerances, support zones, material gradients, and design intent.
- End-to-end platform integration, linking design, simulation, process planning, and monitoring under a unified digital backbone.

Data-driven ML has revolutionized the field of DfAM by enabling the creation of innovative and manufacturable designs. The integration of restrictive and opportunistic DfAM strategies is essential for balancing manufacturability and design freedom. As demonstrated by the application of ML in DfAM has the potential to significantly improve design efficiency, performance, and manufacturability. However, challenges such as data availability and model robustness need to be addressed to fully realize the potential of ML in DfAM.

References

1 Trovato, M., et al., Machine learning trends in design for additive manufacturing, in *Lecture Notes in Mechanical Engineering*. 2024.

2 Venugopal, V. and Anand, S., *Structural and thermal generative design using reinforcement learning-based search strategy for additive manufacturing*. Manufacturing Letters, 2023. **35**: pp. 564–575.

3 Trovato, M., et al., *Machine learning in design for additive manufacturing: a state-of-the-art discussion for a support tool in product design lifecycle*. The International Journal of Advanced Manufacturing Technology, 2025. **137**(5): pp. 2157–2180.

4 Trovato, M. and Cicconi, P., *A decision tree approach for an early evaluation of 3D models in design for additive manufacturing*. Procedia CIRP, 2024. **128**: pp. 96–101.

5 Egan, P.F., *Design for additive manufacturing: recent innovations and future directions*. Designs, 2023. **7**(4): p. 83.

6 Saimon, A. I., et al., *Advancing additive manufacturing through deep learning: a comprehensive review of current progress and future challenges*. IISE Transactions, **58**(1), pp. 33–56, 2025

7 Safdar, M., et al., *Engineering of additive manufacturing features for data-driven solutions: sources, techniques, pipelines, and applications*. 2023, Springer Nature.

8 Zhang, B., et al., Design for additive manufacturing constrained topology optimization of liquid-to-air heat exchangers with ML-based turbulent flow considerations. in *Proceedings of ASME 2024 19th International Manufacturing Science and Engineering Conference, MSEC 2024*. 2024.

9 Ko, H., et al., *Machine learning and knowledge graph based design rule construction for additive manufacturing*. Additive Manufacturing, 2021. **37**: p. 101620.

10 Ullah, A., et al., *A machine learning approach for mechanical component design based on topology optimization considering the restrictions of additive manufacturing*. Journal of Manufacturing and Materials Processing, 2024. **8**(5): p. 220.

11 Zhang, Y., et al., *Predictive manufacturability assessment system for laser powder bed fusion based on a hybrid machine learning model*. Additive Manufacturing, 2021. **41**: p. 101946.

12 Haruna, A., Yang, M., and Jiang, P., *Design for additive manufacturing: a three layered conceptual framework for knowledge-based design*. Proceedings of the Institution of Mechanical Engineers, Part B: Journal of Engineering Manufacture, 2023. **237**(9): pp. 1405–1421.

13 Zhang, C. and Zhao, Y.F., *A hybrid deep learning approach for the design of 2D auxetic metamaterials*. Computer Methods in Applied Mechanics and Engineering, 2025. **441**: p. 117972.

14 Weiss, B.M., et al. Data-driven additive manufacturing constraints for topology optimization. in *Proceedings of the ASME Design Engineering Technical Conference*. 2018.

15 Stachg, D. and Telgkamp, J., *Methodical development of data driven quantitative design rules in additive manufacturing*, preprint, HAW Hamburg. 2023.

16 Aljabali, B.A., Parupelli, S.K., and Desai, S., *Generalized design for additive manufacturing (DfAM) expert system using compliance and design rules*. Machines, 2025. **13**(1): p. 29.

17 Yang, S., et al., *Towards an automated decision support system for the identification of additive manufacturing part candidates*. Journal of Intelligent Manufacturing, 2020. **31**(8): pp. 1917–1933.

18 Zhang, C., et al., *Variant design generation and machine learning aided deformation prediction for auxetic metamaterials*. Mechanics of Materials, 2023. **181**: p. 104642.

Index

Printed and bound by CPI Group (UK) Ltd, Croydon, CR0 4YY
17/09/2026
14964259-0001